SpringerBriefs in Molecular Science

SpringerBriefs in Molecular Science present concise summaries of cutting-edge research and practical applications across a wide spectrum of fields centered around chemistry. Featuring compact volumes of 50 to 125 pages, the series covers a range of content from professional to academic. Typical topics might include:

- A timely report of state-of-the-art analytical techniques
- A bridge between new research results, as published in journal articles, and a contextual literature review
- A snapshot of a hot or emerging topic
- An in-depth case study
- A presentation of core concepts that students must understand in order to make independent contributions

Briefs allow authors to present their ideas and readers to absorb them with minimal time investment. Briefs will be published as part of Springer's eBook collection, with millions of users worldwide. In addition, Briefs will be available for individual print and electronic purchase. Briefs are characterized by fast, global electronic dissemination, standard publishing contracts, easy-to-use manuscript preparation and formatting guidelines, and expedited production schedules. Both solicited and unsolicited manuscripts are considered for publication in this series.

Devanshi Magoo · Anju Srivastava · Sriparna Dutta

Practical Organic Chemistry Workbook

Beginner's Guide to Basics

Devanshi Magoo
Department of Chemistry
Hindu College
University of Delhi
New Delhi, India

Anju Srivastava
Department of Chemistry
Hindu College
University of Delhi
New Delhi, India

Sriparna Dutta
Department of Chemistry
Hindu College
University of Delhi
New Delhi, India

ISSN 2191-5407 ISSN 2191-5415 (electronic)
SpringerBriefs in Molecular Science
ISBN 978-3-031-91985-5 ISBN 978-3-031-91986-2 (eBook)
https://doi.org/10.1007/978-3-031-91986-2

© The Editor(s) (if applicable) and The Author(s), under exclusive license to Springer Nature Switzerland AG 2025

This work is subject to copyright. All rights are solely and exclusively licensed by the Publisher, whether the whole or part of the material is concerned, specifically the rights of translation, reprinting, reuse of illustrations, recitation, broadcasting, reproduction on microfilms or in any other physical way, and transmission or information storage and retrieval, electronic adaptation, computer software, or by similar or dissimilar methodology now known or hereafter developed.
The use of general descriptive names, registered names, trademarks, service marks, etc. in this publication does not imply, even in the absence of a specific statement, that such names are exempt from the relevant protective laws and regulations and therefore free for general use.
The publisher, the authors and the editors are safe to assume that the advice and information in this book are believed to be true and accurate at the date of publication. Neither the publisher nor the authors or the editors give a warranty, expressed or implied, with respect to the material contained herein or for any errors or omissions that may have been made. The publisher remains neutral with regard to jurisdictional claims in published maps and institutional affiliations.

This Springer imprint is published by the registered company Springer Nature Switzerland AG
The registered company address is: Gewerbestrasse 11, 6330 Cham, Switzerland

If disposing of this product, please recycle the paper.

To our Families

Preface

> Science may be equated to God-Generative Operative Destructive

This book is designed as a comprehensive guide for students and practitioners seeking to gain practical experience in organic chemistry. By focusing on practical techniques and experimental procedures, it bridges the gap between theoretical knowledge and hands-on application. Each chapter provides clear instructions, helpful tips, and detailed explanations along with summary sheets to ensure a thorough understanding of essential concepts. Further, each chapter is supplemented with specifically designed pre-lab and post-lab questions to encourage critical thinking and problem-solving skills. Whether you're a beginner or advancing your studies, this book will serve as a valuable resource to build confidence and proficiency in the fascinating world of organic chemistry.

New Delhi, India   Devanshi Magoo
                                                     Anju Srivastava
                                                     Sriparna Dutta

Acknowledgements We would like to express our deepest gratitude to all those who have contributed to the completion of this organic practical workbook. Hindu college has been instrumental in providing motivation and an enriching environment for academic growth and we are thankful to the support and resources offered by the institution in bringing this project to fruition. We also express our sincere gratitude to the Star College Scheme of the Department of Biotechnology, Government of India, for their continued support, motivation, and encouragement in our efforts to develop enriched teaching and learning resources for undergraduate students.

It would not have been possible to accomplish this task without the unwavering support, encouragement, and patience of our family members throughout this endeavour.

A special thanks to the laboratory staff and fellow students for their collaboration, feedback, and assistance during the preparation and testing phases of this book. The knowledge and insights shared by our colleagues helped refine the content and ensure its practical relevance.

Finally, we would like to acknowledge all those who have contributed, either directly or indirectly, in shaping this work.

Competing Interests The authors have no competing interests to declare that are relevant to the content of this manuscript.

Contents

1. Calibration of Thermometer 1
2. Determination of the Melting Points of Unknown Organic Compounds 13
3. Effect of Impurities on Melting Point—The Mixed Melting Point Method 27
4. Purification of Organic Compounds by Recrystallization 41
5. Determination of Boiling Point of Liquid Compounds 59
6. Chromatography 73
7. Detection of Extra Elements 95

Viva-Voce Questions with Answers 117
References 121

Chapter 1
Calibration of Thermometer

1.1 Objectives

To calibrate a liquid in glass laboratory thermometer.
To construct a calibration curve for thermometer.

1.2 About the Experiment

This experiment is designed to introduce you to typical melting point apparatus involving a mercury thermometer inserted in a bath liquid (concentrated sulphuric acid or paraffin wax) contained in a Kjeldahl's flask.

You would learn to calibrate a liquid in glass thermometer and simultaneously determine melting point of organic compounds.

1.3 Introduction

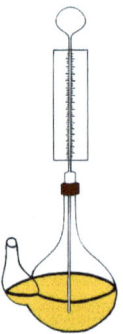

Did you know?

Galileo's Thermoscope

Thermoscope is considered to be the predecessor of thermometer. The first thermoscope was a primitive instrument with an arbitrary scale.

1.3.1 Glass Thermometers

Traditionally, liquid in glass thermometers are the most widely used in industry and laboratory applications. The fill liquids commonly used are mercury or ethanol. Both these liquids have their own set of advantages and disadvantages as shown in Table 1.1. The concept of expansion of a liquid is made use of in these thermometers and they are constructed so that a uniform-diameter capillary tube surmounts a mercury/ ethanol reservoir. The fill liquid is sealed into a narrow glass tube with a small bulb at the bottom. When the temperature is low the liquid does not take up much room and the level in the tube is low. As the temperature rises the liquid expands, and some is pushed out from the bulb into the tube. Since the tube is narrow, even a slight expansion of the liquid results in an appreciable rise in level. The expansion (increase in bulk) is proportional to the increase in temperature, and so the level of the liquid (mercury/ alcohol) in the tube can be used to indicate how warm or cold (i.e. at what temperature) the thermometer and its surroundings are. A scale is marked along the tube dividing it into evenly spaced 'degrees' of temperature. Not all liquids, by any means, expand regularly. Water, for example, is one that behaves in a very unusual way. From 4 °C down to 0 °C it expands as it is cooled.

1.3 Introduction

Table 1.1 Comparison of alcohol and mercury as fill liquids

	Alcohol as fill liquid	Mercury as fill liquid	
Alcohol is colourless so dyestuff is required to be added to make it visible		Mercury is easily visible as its colour is shining silvery	
Its expansion is regular and it has a large temperature coefficient		Its expansion is regular and conducts heat well	
It can measure temperature precisely		It does not wet the walls of the tube	
It clings to the walls of the tube and thread has a tendency to break		Responds quickly to temperature changes and measures temperature precisely	
It is less hazardous and safer to use		Mercury is poisonous and therefore hazardous if tube breaks	
It can measure low temperatures (freezing point − 115 °C)		It cannot be used for measuring very low temperatures as its freezing point is high (−40 °C)	
It cannot be used for measuring high temperatures as its boiling point is low (78 °C)			

1.3.2 Melting Point

Melting point is a characteristic property of a compound that can aid in its identification and purity assessment. It is the temperature at which the solid and liquid phases of a compound are in equilibrium at a pressure of 1 atmosphere. Most organic compounds, when pure melt over a 'sharp and narrow' range of one or two degrees Celsius, hence, melting point (abbreviated as m.p.) is actually a melting range (Fig. 1.1).

The accuracy of melting-point determination can be no better than the accuracy of the thermometer used. Thermometers can give high or low temperature readings of one or two degrees or more, thus there is a need to calibrate the thermometer at hand before starting with any melting point determination. A thermometer should be calibrated by observing the melting points of a series of compounds that are readily available in the pure state and the melting points of which are easily reproducible. For the purpose of calibration, the same apparatus and thermometer should be used for all melting point determinations. It is then convenient and time saving to plot a calibration curve (Fig. 1.2).

Fig. 1.1 Melting point determination using Kjeldahl's flask

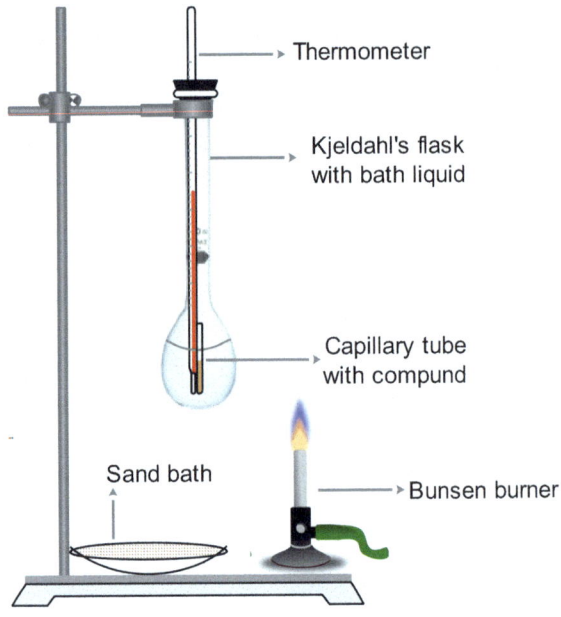

Fig. 1.2 Calibration curve for thermometer

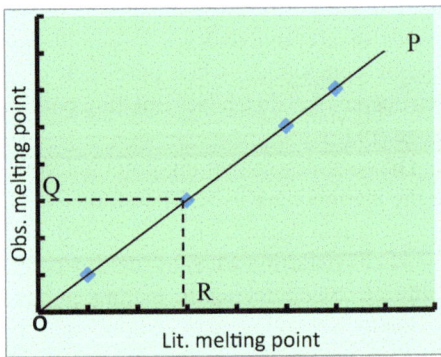

1.4 Pre-lab Questions

1. Why is it necessary to calibrate a thermometer?

   ....
   ....
   ....
   ....
   ....
   ....

1.4 Pre-lab Questions

2. Make a table of all the organic solids used for calibration including their chemical structure, m.p., handling precautions and health hazards.

Organic Compound	Chemical Structure	M. P. (°C)	Handling Precautions	Health Hazards

3. Why do we need to know the melting point of a compound?

....
....
....
....
....
....

4. Name a few commonly used bath liquids for melting point determination. What are their advantages and disadvantages?

....
....
....
....
....
....

5. What are the different scales used to measure temperature?

....
....
....
....
....

6. What is the difference between a digital thermometer and an analogue thermometer?

....
....
....
....

7. Mark as true or false:
   (a) The space above the mercury column in a thermometer is vacuum.
   (b) Alcohol thermometer is suitable for measuring the temperature of boiling water.

1.5 Experimental Section

1.5.1 Chemicals and Apparatus/equipment Required

Chemicals	Apparatus/Equipment
p-Nitrotoluene, m-Dinitrobenzene, Benzoic acid, Salicylic acid, Succinic acid, Anthracene	Kjeldahl's flask with bath liquid, thermometer, capillary tubes, Bunsen burner, sand bath, porous plate, spatula, glass tube

1.5.1.1 Thermometer Calibration Curve

To plot a calibration curve for the given thermometer take the melting points of all the given compounds with the same thermometer and experimental set up. Record the observed melting points and also take note of the literature melting points for all the given organic compounds. Then plot the result with the literature melting point on the x-axis versus observed melting point on the y-axis as shown in Fig. 1.2. Draw a curve OP through these points. Project the observed value Q horizontally to the curve and then vertically down to get the corrected value R for all future determinations. Usually these plots are linear.

1.5.2 Experimental Procedure

1.5.2.1 Sample Preparation

Place a small amount (0.1 g) of the organic compound onto a porous plate or a watch glass. Crush the compound to a fine powder with the help of flat end of the spatula. Take a thin walled capillary tube about 1 mm in diameter and seal it at one end by carefully heating it in the flame using a Bunsen burner. Transfer a small amount of the powdered compound into the capillary tube through its open end. Pack the sample tightly by tapping the capillary tube on the desk holding it between the thumb and index finger or by dropping it through a glass tube. The sample should be 2–3 mm in height.

1.5.2.2 Melting Point Determination

Take the thermometer to be calibrated and insert its upper end through a rubber cork. Attach the capillary tube containing the sample to the lower end of the thermometer that has been previously wetted by the bath liquid. Slide the capillary along the thermometer such that the sample is at the level of the middle of the mercury bulb. The surface tension of the bath liquid holds the moistened capillary in position. Insert the thermometer with the attached capillary into the Kjeldahl's flask dipping the bulb and the capillary into the bath liquid while keeping the open end of the capillary well above the liquid level. Place a sand bath underneath the Kjeldahl's flask as a safety precaution (Fig. 1.1). Heat the Kjeldahl's flask moderately with the Bunsen burner. Observe the temperature at which the sample compound begins to liquefy and at which the solid has completely disappeared. For a pure compound this temperature range should be small (1–2 °C). Record the temperature and allow the bath to cool before determining the melting point for the next compound. Similarly determine the melting points for all the given organic compounds and record the observations.

Melting point determination using electrically heated apparatus and BODMEL method have been discussed in Unit 2.

1.6 Post-lab Questions

1. How do you properly cool off a melting point thermometer?

2. For the purpose of calibration, the same apparatus and thermometer should be used for all melting point determinations. Why?

3. Does the rate of heating influence the melting point?

4. How is mercury prevented from contracting into the glass bulb immediately after melting point determination?

5. Why is it necessary to make a compact packing of the sample in the capillary tube while determining its melting point?

6. The melting point of a compound which has already been melted once often differs from the original value. Why?

1.7 Summary Sheet

Thermometer Calibration
Observation Table:

S.No.	Organic Compound	Obs. m.p. (°C)	Lit. m.p. (°C)	Temp. Correction [Obs. m.p - Lit. m.p.] (°C)
1.	*p*-Nitrotoluene			
2.	*m*-Dinitrobenzene			
3.	Benzoic acid			
4.	Salicylic acid			
5.	Succinic acid			
6.	Anthracene			

Calibration curve

Literature melting point (x) versus observed melting (y):

Results & Discussion:……………………………………………………….
………………………………………………………………………………...……

Notes

1.7 Summary Sheet

Chapter 2
Determination of the Melting Points of Unknown Organic Compounds

2.1 Objectives

To determine the melting point and identify the given unknown organic compound.
To understand the use of electrically heated melting point apparatus.
To understand BODMEL method of melting point determination.

2.2 About the Experiment

This experiment is essentially an extension of Experiment 1 where you had learnt the technique of melting point determination by Kjeldahl's method. You would now be introduced to BODMEL apparatus and an electrically heated melting point apparatus to determine the melting point of unknown organic compounds. This experiment would be a prelude to Experiment 3 where you would learn to identify unknown organic compounds by mixed melting point method.

2.3 Introduction

With the advent of spectrophotometric methods, IR (infrared), UV (ultra-violet), NMR (nuclear magnetic resonance) spectra can give exact details of the identity of an unknown compound; however, determination of melting point of a compound is still in an inseparable part of compound identification. The melting point, more appropriately called as the melting range of a pure organic compound is the temperature range at which the solid exists in equilibrium with its liquid state under an external pressure of one atmosphere. A solid organic compound possesses an orderly crystalline

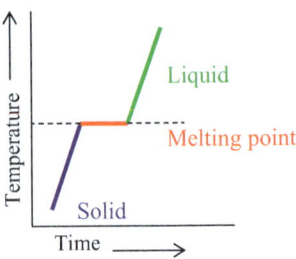

Fig. 2.1 Melting range for a compound

lattice and as heat is supplied, its molecules acquire enough energy to overcome the intermolecular forces binding them together eventually changing the solid to a liquid.

The process of melting is not instant as it takes some time for the heat to transfer from the heating source to the molecules of the organic sample that then absorb the energy to physically break the binding forces. Thus, the compound melts over a range of temperature which is defined as the span of temperature or the small difference observed between the temperature at which the crystals first begin to liquefy to the temperature at which the entire sample is liquid (Fig. 2.1). Most pure organic compounds melt within a narrow temperature range of within 1–2 °C, however when an organic compound is impure, its melting point is lowered and a broadened range is observed (>3 °C).

Determining the melting range of a compound is useful for two purposes:

1. It helps in assessing the purity of the compound by comparing its melting range with the known range for a pure authentic sample of the same compound. A depressed and broadened range would indicate that the sample is impure or contaminated.
2. It helps in identification of an unknown sample by comparing its observed melting range with that of known compounds.

2.3.1 Melting Point Determination by BODMEL Method

The past decades have witnessed a seismic shift in the approach towards organic analysis. Macro scale experiments have given way to small scale/micro scale methods of analysis owing to tremendous environmental and cost benefits. However, transitioning towards micro requires the use of appropriately sized apparatus such as miniaturized versions of standard apparatus or small scale kits. For the determination of melting point, a new piece of glass apparatus known as the "BODMEL Apparatus" has been designed, fabricated and tested which is safe to use, offers ease of set-up, gives reliable and quick results and can be used for more than one function.

Figure 2.2 displays the set-up of the BODMEL's apparatus which comprises of a central tube with a constriction and two finger-like projections close to the neck. The main tube is covered with an outer jacket which plays a crucial role of serving as an

2.3 Introduction

Fig. 2.2 BODMEL's apparatus

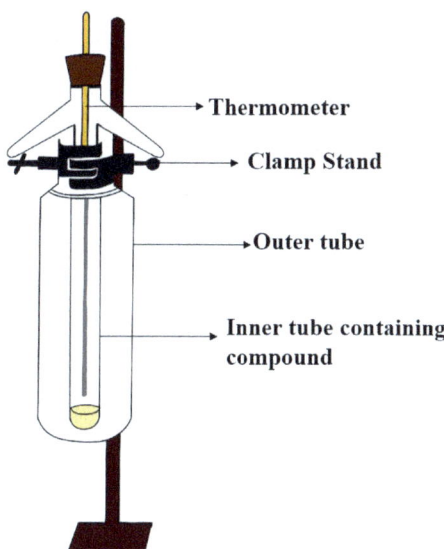

effective air bath having two distinctive holes for atmospheric contact. Notably, it is a single apparatus which does not have ground glass joints.

2.3.2 Using the Electrically Heated Melting Point Apparatus

You have already learnt the use of Kjeldahl's method for determination of melting point in Experiment 1. A more modern method for finding out the melting point involves the use of electrically heated melting point apparatus also called the mel-temp apparatus. With such an apparatus high temperatures can be reached safely avoiding the risks posed by the use of sulphuric acid or hot oil baths used in Kjeldahl's or Thiele's melting point set-up. Figure 2.3 shows a typical model of the mel-temp apparatus which consists of the following components:

1. The base accomodating the transformer and the main controls.
2. The upper panel housing the optical system for illuminating and viewing the samples.
3. A block to hold the thermometer and upto three samples. While investigating just one sample, the other two capillary inlets in the block should be occupied with empty tubes.

Depending upon whether the melting point of sample under investigation is known or unknown, slightly different methods can be adopted to have an accurate measurement of the melting range. If the melting point of the sample is unknown, first step will be to carry out a preliminary run by employing a rapid rate of heating throughout.

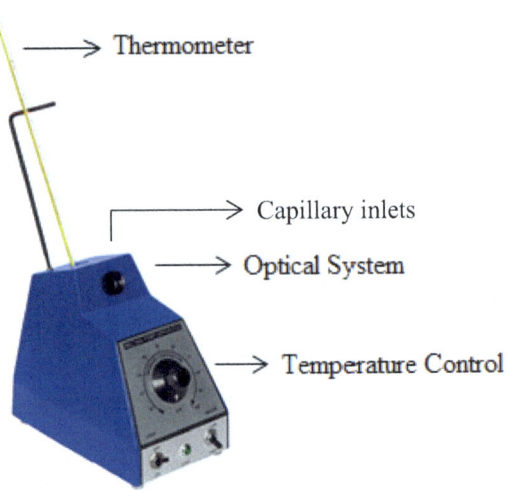

Fig. 2.3 Electrically heated melting point apparatus

This would give an approximate melting range for the compound. After establishing the approximate range a more careful measurement can be made in the final run employing a slower rate of heating.

2.3.2.1 Measuring an Approximate Melting Point

For a compound of which the melting point is unknown, it is advisable to first establish an approximate melting point. An approximate melting point can be measured using a rate of 5–10 °C /minute rise in temperature. The following protocol should be adopted:

1. Place the bulb end of the thermometer in the space provided in the block.
2. Switch on the apparatus using the on/off switch.
3. Insert the melting point capillary tube containing the sample into the inlet. Place empty capillary tubes in the rest of the two inlets.
4. Turn the temperature control knob to achieve a rate of temperature rise of about 5–10 °C/minute.
5. Observe the approximate melting range of the sample by looking through the magnifying lens of the apparatus. In case the rate of rise in temperature decreases to less than 5 °C/minute, turn the temperature control knob to a higher value to increase the rate.
6. When finished, bring the temperature control knob to "0" or turn the unit off.

2.3.2.2 Measuring a Precise Melting Point

Once an estimate of the approximate melting point has been made or in case the melting point of the compound is already known from the literature, the apparatus

2.3 Introduction

can be utilized to find the precise melting range for the substance by using a rate of 1–2 °C /minute rise in temperature. The following protocol should be adopted:

1. Place the bulb end of the thermometer in the space provided in the block.
2. Switch on the apparatus using the on/off switch.
3. Insert the melting point capillary tube containing the sample into the inlet. Place empty capillary tubes in the rest of the two inlets.
4. Turn the temperature control knob so as to initially achieve a rate of 5–10 °C/minute till the block has reached a temperature of about 20 °C below the approximate melting point measured (or for the known compound, about 20 °C below the literature value).
5. Now adjust the temperature control knob so that the rate of rise in temperature is 1–2 °C/ minute which would result in a more careful and precise measurement of the melting point.
6. Observe the precise melting range of the sample by looking through the magnifying lens of the apparatus carefully maintaining the rate of temperature rise within 1–2 °C/ minute.
7. When finished, bring the temperature control knob to "0" or turn the unit off.

2.3.2.3 Stages of Melting

While monitoring the sample through the illuminated magnifying lens of the apparatus, you will be able to observe the four stages of melting (Fig. 2.4):

a. First indication of change may be shrinking or shriveling.
b. First sign of liquid formation, formation of a droplet. Lower limit of the melting range should be recorded at this point.
c. Formation of a meniscus.

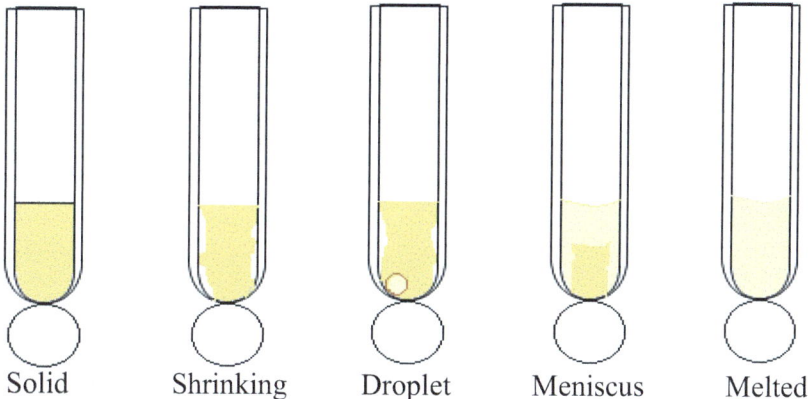

Fig. 2.4 Four stages of melting for a solid compound

d. Complete melting and formation of clear liquid. Upper limit of the melting range should be recorded at this point.

Not all compounds will behave in this ideal manner and may undergo unusual behavior before melting. For instance, some compounds decompose on melting and it may be characterized by discolouration of the sample. Such decomposition points are usually indicated with a symbol of "*d*" after the temperature mentioned in the table of physical properties. For example, 2,4,6-Trinitrobenzoic acid, m.p. 220 °C (*d*). It is also typical of some compounds to shrivel or soften just before melting which is usually due to a change in the crystal structure. This is also called sintering. Certain compounds tend to "sweat" or release solvent of crystallization before melting. Also, there are solids with high vapour pressure that sublime at or below their melting points. Such a behavior is represented by "s" or "sub" or "subl" against the sublimation temperature, for example, Fumaric acid: 200 °C (*subl*).

You would be able to recognize and distinguish melting from shrinkage, softening, decomposition etc. by practice and experience; however, at this stage observing the four main stages described would make you have a reasonably good understanding of the melting range for a compound. You should be observant of the temperature range between the first sign of liquid formation and formation of a clear liquid.

2.4 Pre-lab Questions

1. Why melting range is a more appropriate term than melting point?

   ....
   ....
   ....
   ....
   ....
   ....

2. What are the advantages of an electrically heated melting point apparatus over other traditional methods?

   ....
   ....
   ....
   ....
   ....
   ....

2.4 Pre-lab Questions

3. Heating the sample too quickly in the region of the melting point would result in a higher experimentally determined value than the actual value. Explain why?

   ....
   ....
   ....
   ....
   ....
   ....

4. You have a sample that you are sure is salicylic acid, which has a literature melting range of 157–158 °C.

   a. Suppose you run your sample and observe a melting range of 157–163 °C. Is your sample impure, or did you heat too fast?

   ....
   ....

   b. Suppose you run your sample and observe a melting range of 151–157 °C. Is your sample impure, or did you heat too fast?

   ....
   ....

5. A pure organic compound has a _____ melting range while an impure compound as a _____ melting range.
6. What is sintering?

   ....
   ....
   ....
   ....
   ....
   ....

7. What are the stages of melting that one needs to be vigilant about while determining the melting range of a compound?

   ....
   ....
   ....
   ....
   ....
   ....

2.5 Experimental Section

2.5.1 *Chemicals and Apparatus/equipment Required:*

Chemicals	Apparatus/Equipment
Coded unknown sample Known organic compounds*: *m*-dinitrobenzene, benzoic acid, urea, salicylic acid, 3,5-dinitrobenzoic acid	Bodmel Apparatus/Mel-temp apparatus/ Kjeldahl's melting point set up, Bunsen burner Thermometer, capillary tubes, porous plate, spatula

*You may be simply provided with a list of known organic compounds with their literature melting range to compare with the m.p. of your unknown sample or you may be provided with known compounds for each of which you will have to determine the melting range prior to determining the melting range of the unknown sample given to you.

2.5.2 *Experimental Procedure*

2.5.2.1 Sample Preparation

To record the melting point range of your unknown sample, place a small amount of the sample on a porous plate and crush it to a fine powder with the help of flat end of the spatula. Prepare two capillary tubes (use sealed tubes provided with the mel-temp apparatus or capillaries can be sealed using the procedure explained in Experiment 1) by filling each to a height of 2–3 mm with the powdered sample. Pack the sample tightly by tapping the capillary tube on the desk holding it between the thumb and index finger or by dropping it through a glass tube.

2.5.2.2 Melting Point Determination

Before beginning with the melting point determination, be sure that you are using a calibrated thermometer. If not calibrate the thermometer using the procedure you learnt in Experiment 1.

Melting Point Determination using BODMEL's METHOD

In order to determine the melting point using this apparatus, firstly with the help of a long dropper, approximately 2 mL concentrated H_2SO_4 is taken in the main tube. Thereafter, the thermometer and capillary filled with compound are stuck together using the adhesive property of H_2SO_4. Similar to the conventional methods, this assembly is carefully lowered into the acid bath, cork is secured and heating is accomplished in a uniform manner. Melting points are obtained smoothly.

2.5 Experimental Section

Melting Point Determination using Electric Apparatus

Place one of the prepared capillary tubes in the inlet provided in the mel-temp apparatus and set the apparatus controls to raise the temperature at a relatively fast rate, about 10 °C/minute. Record the temperature between the first visible evidence of liquid formation (the sample may appear a little moist, or a tiny droplet of liquid may be observed) to the complete liquefaction of the sample. This would give you an idea of the approximate melting range.

Once the approximate range has been determined, allow the melting point apparatus to cool to about 20 °C below the temperature at which the sample just began to melt. Then insert the second capillary tube and set the apparatus controls to let the temperature rise more slowly this time, at the rate of about 1–2 °C/min.

Record carefully, the temperatures at which the sample begins to melt, and at which the sample has completely melted. This gives the precise melting rage of the unknown sample. Use this range to compare with the melting ranges of list of known compounds provided to identify the unknown sample.

Sometimes, there may be a possibility when more than one compound lies in the melting range of the unknown sample. In such a case mixed melting point method is a more appropriate way to find the identity of the unknown sample. You will learn more about it in Experiment 4.

2.5.2.3 Some Useful Pointers for a Precise Melting Point Determination

- Always make sure you are using a dry sample while determining the melting point. Water or solvent present in a damp sample act as impurities thus giving erroneous result.
- Use a small amount of sample in the capillary tube. Over-filling the tube might cause uneven heating of the sample resulting in broader ranges and possibly a false indication of impurity.
- Pack your sample tight. Loosely packed sample will also cause uneven heating.
- For unknown samples it is always advisable to measure an approximate melting range before finding out the precise result.
- A sample once meted should be discarded as it may have oxidized, decomposed or rearranged during the process heating and cooling.

2.6 Post-lab Questions

1. Make a table of all the known organic solids being used for the identification of the unknown compound including their chemical structure, literature m.p., handling precautions and health hazards.

Organic Compound	Chemical Structure	Literature M. P.	Handling Precautions	Health Hazards

2. Why is it wise to fill a relatively small amount of sample as opposed to filling in lots of sample in the melting point capillary? How will the melting range be perturbed by putting in large amount of sample?

....
....
....
....
....
....

3. Why is it necessary to have the sample completely dried before taking the melting point?

....
....
....
....
....
....

2.6 Post-lab Questions

4. How is having a finely powdered sample advantageous over a coarse sample for determining the melting range? What effect would a coarse sample have on the melting range?

   ....
   ....
   ....
   ....
   ....
   ....

5. Suppose you isolated the following compounds in lab and observed the melting range given in the table below. For each one of them, look up the literature melting range and judge the purity of the compound.

Compound	Observed melting point
Oxalic acid	180–182 °C
Naphthalene	79–80 °C
Benzophenone	44–46 °C
Cinnamic acid	130–132 °C
α-naphthol	90–92 °C

Note for instructor Provide known compounds with significant gap in melting ranges as the experiment focuses on unknown identification by comparison.

2.7 Summary Sheet

Melting range for given Known Compounds:

S. No.	Known Compound	Observed/ Literature m.p. (°C)

Identification of the Unknown Compound:

Unknown Sample Code: _____

  Observed melting range of unknown sample: _____

– On comparison with the melting range of known compounds, the given unknown sample is found to be: _____

  The structure of the identified compound is:

Results and Discussion

....

....

Notes

Chapter 3
Effect of Impurities on Melting Point—The Mixed Melting Point Method

3.1 Objectives

To understand the effect of impurities on the melting range of organic compounds. To get familiar with the mixed melting point method for identification of unknown compounds.

3.2 About the Experiment

By this time you have a fair grasp over the technique of melting point determination. This experiment is a level up where you will be using the technique for identification of unknown compounds by the mixed melting point method while simultaneously building an understanding of the effects that impurities have on the melting range of organic compounds.

3.3 Introduction

3.3.1 Melting Range

Let us delve deeper into the theory of melting point determination by pondering over the appropriateness of the term "melting range". It is commonly stated 'an impure solid melts at a lower temperature and over a wider range'. Here, by the term impure solid, we mean that the solid is mixed with one or more than one impurity. In other words we can call it a mixture. Take the example of butter which is a mixture of different types of triglycerides. We can define its stages of melting as for any organic compound. It is a solid when stored at 5 °C, at room temperature (~25 °C) it softens and liquefies completely at temperature beyond 40 °C. Thus, this solid mixture of organic compounds melts over a wide temperature range. Presence of even small amounts of impurities lowers and broadens the melting temperature range. Such a behavior of solids is unfailing and is thus a dependable test to judge the purity of a compound.

3.3.2 Effect of Impurities on Melting Range

Due to presence of impurities the lower end of the melting range drops resulting in an overall depression of the melting range. Further, the lower end often drops to a greater extent than the upper end leading to broadening of the range.

The reason for this depression and broadening of the melting range for an impure compound lies in the fact that the impurities being different in structure do not fit correctly into the lattice structure of the compound causing disruption in the intermolecular attractions and organization of the lattice at the molecular level. The typical result of such a disorder is twofold (Fig. 3.1):

a. Depression of melting range

The disorder causes weakening of the lattice so that it requires lesser energy and can be easily broken down. In other words, the disorderly structure melts more easily at a reduced temperature thus causing a depression in the melting range.

b. Broadening of melting range

The disordered lattice structure no more remains uniform so that the molecules lying closest to the impurity melt faster. Whereas, the relatively undisturbed lattice lying further away from the impurity melts at or close to the normal melting temperature. This effect manifests itself as broadening of the melting range.

Figure 3.2 shows the disruption of lattice structure of a compound due to incorporation of impurity.

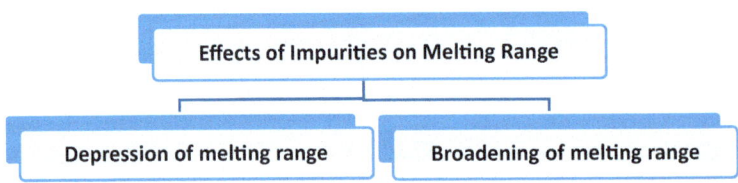

Fig. 3.1 Two-fold effect of impurities on melting range

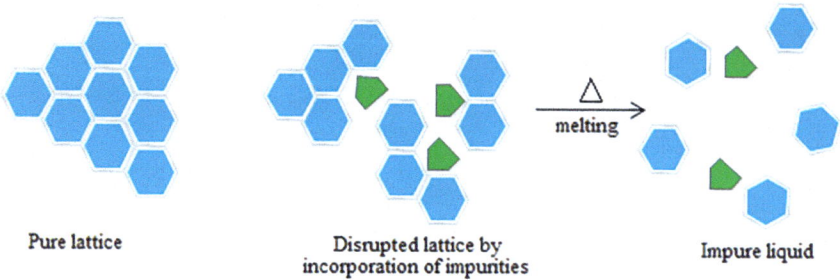

Fig. 3.2 Disruption of lattice structure due to impurities

3.3.3 More on How Impurities Effect the Melting Range

- Impure solids are often purified by recrystallization discussed in experiment 4 followed by removal of solvent by filtration. It is important to allow the sample to get completely dried as any remaining solvent would act as a contaminant resulting in depressing/broadening of the melting range.
- An insoluble impurity like a small piece of wood will have a negligible or no effect on the melting range of a compound as only a few molecules in close proximity will get affected. Whereas a marked effect will be observed in the presence of soluble impurities which have the capability of getting incorporated into the lattice structure and disrupting it at the molecular level.
- The melting point of a compound can never be raised by the incorporation of an impurity, it can only be depressed. For example, if a compound A melts at 120 °C and is mixed with small amount of a compound B (impurity) having a higher melting point of 195 °C, then the resulting mixture will begin to melt a temperature lower than 120 °C i.e. the melting point of the major component will be depressed.
- The melting point for a mixture having two pure compounds with exactly the same melting points will be depressed due to one acting as impurity for the other.

3.3.4 The Mixed Melting Points

The technique of mixed melting points has evolved taking advantage of the fact that mixtures exhibit a depression in melting point. By using this method it can be verified whether or not an unknown compound is identical to the suspected known compound. If two pure organic solids are mixed together thoroughly, either of the two possibilities may arise:

1. If the two are identical, the melting point will remain unaffected.
2. If the two are different, each will act as an impurity in the other, and the melting range will be noticeably depressed.

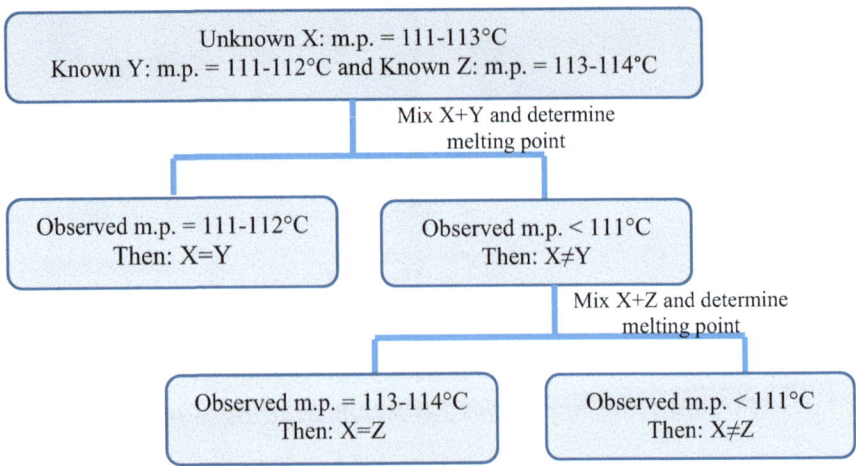

Fig. 3.3 Identification of unknown by mixed melting point method

Consider an example where X is an unknown compound with a melting point of 111–113 °C and appears identical to known compounds Y (m.p. = 111–112 °C) and Z (m.p. = 113–114 °C). Then X may be identical to Y or Z as the melting ranges lie very close to one another. To identify X mixture melting points of the two mixtures X + Y and X + Z can be determined. If the mixture melting point of X + Y is not depressed then, X = Y; and if the mixture melting point of X + Z is not depressed then, X = Z. However if both the mixture melting points are depressed then X ≠ Y or Z. The following flow diagram (Fig. 3.3) explains the protocol of mixed melting point method for identification of unknown compounds:

Let us consider another example with one of the most common organic compounds you have been using in lab, benzoic acid. Suppose, in your organic chemistry lab you find an un-labeled bottle of a compound which you suspect to be benzoic acid. How will you ensure whether or not the compound in the bottle is benzoic acid? First you would determine the melting point of the unknown compound and let us suppose you find it to be 119–120 °C which is close to the actual melting point of benzoic acid. So the unknown compound could very well be benzoic acid. To alleviate any doubt, the mixed melting point method comes in use. For this, you would obtain a genuine sample of benzoic acid from the lab stockroom and mix a small amount of this pure sample with the suspected "unknown" compound and then determine the melting point of the mixture. If the melting point of the mixture so formed is still 119–120 °C, it would be certain that the unknown compound was benzoic acid—all that you have done is to mix benzoic acid with benzoic acid, so that the melting point remains unaffected. If the "unknown" was not benzoic acid, then the benzoic acid that you added would have acted as an impurity, and the unknown would have melted over a lower and a broader range of temperature. Thus if on mixing two compounds, the melting range remains unchanged, then the two compounds are identical. But if they are different, the melting range would surely be lowered and broadened.

3.3.5 Exceptions

- *It should be noted that there is a unique possibility for a mixture of two different compounds to give a sharp melting point. Such a mixture contains the two compounds in a specific ratio and has a lower melting point than any other composition of the mixture of the two compounds. This typical mixture is called the **eutectic mixture** and its melting point is called the **eutectic point**. Thus, a eutectic mixture could be mistaken for a pure compound. However, if the composition of the eutectic mixture is changed by adding a small amount of either of the two compounds (assuming they are both known), the melting point of the resultant mixture will be **higher** and more spread out than the melting point of the eutectic mixture.*
- *There may be instances where the melting points of some mixtures may be higher than the individual components. Such a situation is possible if two compounds of the mixture form solid solutions due to complete solubility in the solid state or they may form an addition compound of higher melting point.*

Thus, in its own merit the mixed melting point method has a remarkable practical application but it does have a few exceptions which should be kept in mind while making use of it.

3.4 Pre-lab Questions

1. Define the following terms:

    a. Melting range

    ....
    ....

    b. Eutectic mixture

    ....
    ....

2. How does the melting range of a compound get perturbed due to the presence of impurities?

    ....
    ....
    ....
    ....
    ....
    ....

3. Why does an impure compound exhibit a depressed melting range?

4. How does the presence of impurity result in broadening of the melting range of a compound?

5. Are there any exceptions which question the infallibility of mixed melting point method?

6. What is the effect of an insoluble impurity, such as sodium sulfate, on the observed melting point of a compound?

3.5 Experimental Section

3.5.1 *Chemicals and Apparatus/Equipment Required*

Chemicals	Apparatus/Equipment
Coded unknown samples Known organic compounds: Benzamide, phthalic anhydride, benzoic acid, succinic anhydride, 2-naphthol	Mel-temp apparatus/Kjeldahl's melting point set up, Bunsen burner Thermometer, capillary tubes, porous plate, spatula

3.5.2 *Experimental Procedure*

3.5.2.1 Observing the Depression in Melting Range

In this experiment you would be given two known compounds of which the melting ranges lie close to one another. You will have to record the melting point of each of these individually and then by taking a mixture of the two. Performing this experiment would help you observe how mixtures exhibit a depression in melting range.

Single melting point determination

Procure authentic samples of benzoic acid and 2-naphthol from your instructor. Place a small amount of benzoic acid on a porous plate and crush it to a fine powder with the help of flat end of the spatula. Prepare a tightly packed capillary tube by filling it to a height of 2–3 mm with the powdered benzoic acid. Likewise prepare a capillary for 2-naphthol. Determine and record the melting range of each of the two compounds using the mel-temp apparatus or Kjeldahl's melting point set up/BODMEL set-u[.

Mixed melting point determination

Place approximately equal amounts of benzoic acid and 2-naphthol on a clean porous plate. Grind the two compounds together and prepare an intimate mixture. Prepare three capillary tubes, one filled with finely powdered benzoic acid, second with finely powdered 2-naphthol and the third capillary with the mixture of the two compounds. Place the three capillaries in the inlets provided in the mel-temp apparatus simultaneously. (In case mel-temp apparatus is not available, individual melting point determination can be carried out using Kjeldahl's melting point set up/BODMEL set-up). Carefully observe the melting behavior of the pure compounds and the depression showed by the mixture. Record the melting points observed for each of the three.

Note: Same exercise can be done taking urea and cinnamic acid or 3-chlorobenzoic acid and salicylic acid.

Table 3.1 Series of known compounds for the given unknowns

Unknown 1		Unknown 2	
Series I	Benzamide	Series II	Benzoic acid
	Phthalic anhydride		Succinic anhydride

3.5.2.2 Identification of Unknowns by the Mixed Melting Point Method

You will be given two unknown compounds (Unknown 1, Unknown 2) each belonging to two different series (I, II) of known organic compounds given in Table 3.1. Unknown 1 will be one of the compounds of Series I. Similarly Unknown 2 will be one of the compounds of Series II. Using the mixed melting point method you will have to identify the unknown compounds.

Melting point determination

Determine the melting point of Unknown 1, benzamide and phthalic anhydride individually. To identify the unknown 1, prepare two mixtures and determine their melting ranges:
  Mixture 1 = (Unknown 1 + benzamide)
  Mixture 2 = (Unknown 1 + phthalic anhydride)
  If the melting point of Mixture 1 is depressed, then Unknown 1 $\neq$ benzamide.
  If the melting point of Mixture 2 is depressed, then Unknown 2 $\neq$ phthalic anhydride.
  Similarly perform the experiment for identification of Unknown 2.

3.6 Post-lab Questions

1. Make a table of all the known organic solids being used for the identification of the unknown compound including their chemical structure, literature m.p., handling precautions and health hazards.

....
....
....
....
....
....

3.6 Post-lab Questions

2. How can mixed melting point method be used for identification of an unknown organic compound?

   ....
   ....
   ....
   ....
   ....
   ....

3. A student was given a white solid for an unknown. Its melting point range was 132–133 °C. The student has previously worked with urea, and had observed that it was a white crystalline solid with a melting point of 132–133 °C.

   (a) Can the student conclude on these observations that the unknown is benzoic acid. Why or why not?

   ....
   ....
   ....
   ....
   ....
   ....

   (b) What additional experimental work should be done to verify this compound?

   ....
   ....
   ....
   ....
   ....
   ....

4. You and your lab partner synthesize m-dinitrobenzene in lab and recrystallize it before taking the melting point. You observe a melting range of 84–90 °C, while your partner observes a value of 88–90 °C. Explain how you can get two different values with exactly the same sample.

   ....
   ....
   ....
   ....
   ....
   ....

5. An unknown sample X melts at 148–152 °C, and is thought to be either candidate A (known range is 141–142 °C) or B (known range is 154–155 °C). Is sample X identical to candidate A, or is it an impure and thus depressed version of candidate B. Give reasons for your answer.

   ....
   ....
   ....
   ....
   ....
   ....

6. What would be the effect of adding salt on the melting point of ice? Explain.

   ....
   ....
   ....
   ....
   ....
   ....

7. Formulate a flowchart similar to the one shown in Fig. 4.3 to show how you identified your unknown 2.

3.7 Summary Sheet

A. Observing the depression in melting range

Observed melting range of benzoic acid: _____

Observed melting range of 2-naphthol: _____

Observed melting range of mixture: _____

B. Identification of unknowns by the mixed melting point method

Identification of Unknown 1

Sample	Observed melting range (°C)
Unknown 1	
Benzamide	
Phthalic anhydride	
Unknown 1+ Benzamide	
Unknown 1+ Phthalic anhydride	

Identification of Unknown 2

Sample	Observed melting range (°C)
Unknown 2	
Benzoic acid	
Succinic anhydride	
Unknown 2+ Benzoic acid	
Unknown 2+ Succinic anhydride	

Results and Discussion

....

....

Notes

3.7 Summary Sheet

Chapter 4
Purification of Organic Compounds by Recrystallization

4.1 Objectives

To purify the given organic compound(s) by recrystallization method.
To understand the use of single solvent and mixed solvent systems for recrystallization.

4.2 About the Experiment

This experiment is designed to show how organic compounds (solids) can be purified by recrystallization using different solvents. You would learn about the solubility of organic compounds and thus the choice of solvent for recrystallization. Different techniques you would learn through this experiment include hot filtration and vacuum filtration. You would also learn the use of activated charcoal and how to make a fluted filter paper.

4.3 Introduction

4.3.1 The Need for Purification

Organic compounds, whether isolated from natural sources or synthetically prepared are often contaminated with impurities. Compounds isolated from organic reactions may have impurities in the form of small amounts of unreacted starting materials or by-products obtained along with the desired product. There are many reasons why we may need to purify an organic compound. For instance, a higher level of purity may be needed for further use in an organic synthesis or for final characterization, especially if the compound is new or unknown. Especially, in medicine, an organic compound must be of very high purity before it can be administered in the body as a drug. As we learned in the previous experiment, one way we can determine if a compound is pure is to measure the melting point. A pure compound has a sharp and narrow range of melting point, while an impure compound has a broad and depressed melting point.

> **Did You Know?**
>
> **Crystallization versus Recrystallization**
>
> Crystallization is a separation technique involving precipitation of crystals from a solution owing to changes in solubility conditions of the solute in the solution. Recrystallization is a technique employed to purify the crystals obtained from crystallization method. Though crystallization separates the compound in almost pure form, however, when the crystals form some of the impurities may trap in it. By recrystallization method, these impurities can be removed to a greater extent.

It is therefore imperative to adopt purification methods so as to obtain the desired product in the pure form. Techniques employed for purification of liquid organic compounds include simple distillation, fractional distillation, vacuum distillation, steam distillation, solvent extraction etc. Organic solids on the other hand are usually purified by the methods of crystallization, fractional crystallization, sublimation, solvent extraction etc. Chromatography is another effective method for separation and purification of compounds about which you would learn in Unit 6.

4.3.2 Principle of Recrystallization

Solid organic compounds are usually purified by the process of recrystallization from a suitable solvent or a mixture of solvents. This process of purification is based on differential solubility of solid organic compounds and their impurities in a given solvent. The method involves finding a suitable solvent that can dissolve the impure/crude compound when hot and precipitate out the pure compound on cooling.

Certain assumptions that are considered in the underlying principle are that the impurities are more soluble than the actual compound and that they are present in such small proportions, that in spite of their comparative solubilities, they can be eliminated by recrystallization i.e. the solution will get saturated with respect to the pure compound which is much larger in quantity and not with respect to the impurities which thus remain dissolved in the solvent.

The process of recrystallization essentially involves the following steps:

1. Choosing a suitable solvent or mixture of solvents to dissolve the impure compound.
2. Dissolving the impure compound in the selected solvent at or near its boiling point.
3. Decolourising with charcoal (activated carbon) to remove coloured impurities, if present.
4. Filtering the hot solution to remove insoluble suspended impurities.
5. Allowing the hot solution to cool gradually to effect crystallization of the pure compound.
6. Separating the crop of crystals by filtration.
7. Drying the crystals and testing for purity, usually by melting point determination.

The above process may be repeated to achieve the desired degree of purity often confirmed by unchanged melting point and also by spectroscopic methods.

Did You Know?

Decolourising Charcoal (activated charcoal)

For obtaining activated charcoal pyrolysis of carbonaceous substances like wood is carried out at a very high temperature to extract pure carbon (carbonization). Once this porous carbon is produced it needs to undergo oxidation by gas or chemical treatment to improve its adsorbent properties. This activated charcoal can remove coloured impurities and resinous matter from many solutions. It does so by the process of adsorption, by attracting these molecules to its surface.

4.3.3 Nature of Impurities and Choice of Solvent

The impurities present in an organic compound to be recrystallized may be classified as soluble, insoluble, and coloured. Insoluble impurities can be removed from a compound fairly easily. The compound to be purified is dissolved in a solvent, the solution is filtered to remove the insoluble impurities, and the solvent is evaporated to produce the solid compound. The insoluble impurities are left behind on the filter paper. Impurities, if coloured can also be removed in a similar way but with an additional step. The compound is dissolved in a solvent and decolourising charcoal is added, the solution is filtered as before, and the solvent is evaporated to produce the solid compound. The charcoal, that has adsorbed the colored impurities, is left behind on the filter paper.

However, soluble impurities cannot be as easily removed as their solubility characteristics are similar to those of the desired compound (hence the name soluble impurity). In order to remove such impurities, a suitable solvent is required to be chosen depending upon the solubility behavior of the compound to be purified.

An organic compound usually exhibits one of three general solubility behaviors:

(a) The compound has a low solubility in both hot and cold solvent.
(b) The compound has a high solubility in both hot and cold solvent.
(c) The compound has a high solubility in hot solvent but a low solubility in cold solvent.

Solvents in which the compound exhibits behavior types (a) and (b) are not useful for recrystallization of that compound. A solvent with respect to which the compound exhibits behavior type (c), i.e., high solubility at high temperatures and low solubility at low temperatures, is the one that is suitable for use as a recrystallization solvent. A graphical representation of the above facts is given in Fig. 4.1.

Recrystallization therefore depends primarily on solubility relationships. Solubility can be regarded as a relative term. The questions, "what solvent would dissolve a solid"? and "how much would it dissolve"? are answered in terms of the degree of solubility which may be defined in terms very soluble, soluble, sparingly soluble and insoluble and expressed in grams per 100 mL of the solvent.

Although theoretical considerations cannot be really put to use in choosing the correct solvent for recrystallization, however some general solubility principles help making the choice easier. For predicting solubilities, the rule of "like dissolves like" is often stated, i.e. a compound would generally dissolve in a given solvent if the forces that hold its own molecules together are similar to the forces that hold the molecules of the solvent together. A polar compound would tend to dissolve in a polar solvent and a nonpolar compound would tend to dissolve in a nonpolar solvent.

Fig. 4.1 Solubility behavior of compound in solvents w.r.t temperature

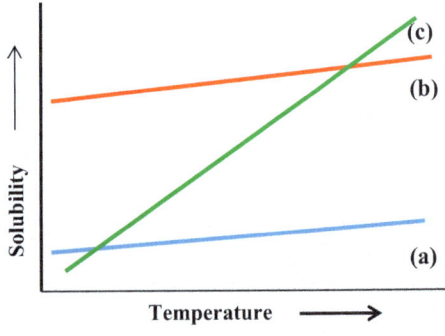

4.3 Introduction

The solubility of an organic compound would also depend on the functional groups that are present in it. Compounds containing groups such as -OH, -NH$_2$, -CONH-, etc. are usually more soluble in polar solvents such as alcohols or water than in hydrocarbons. Whereas, non-polar molecules possessing very few or no polar groups would dissolve in hydrocarbons. Some general features that a solvent suitable for recrystallization should possess are:

1. A good recrystallization solvent should have a high degree of solubility for the compound to be purified at elevated temperatures (at or near boiling point) and low solubility at room temperature or lower temperature.
2. It should not be chemically reactive towards the compound.
3. The solvent should exhibit a differential behavior towards the solubility of the compound and the impurities, i.e. it should either dissolve the impurities very readily or to a very small extent.
4. The boiling point of the solvent should be lower than the melting point of the compound being recrystallized.
5. The solvent should be easily removable from the purified compound.
6. It should pose minimum or no environmental/health hazard.

> *Did You Know?*
>
> *Solubility & Crystal Lattice Stability*
>
> *Irrespective of the type of compound, the more stable the crystal lattice, the higher the melting point, the less soluble the compound. The case of isomeric nitro benzoic acids is an appropriate example:*
>
> | ortho | meta | para |
> |---|---|---|
> | 147°C | 141°C | 242°C |
> | 28 | 33 | 2.2 |

Some common recrystallization solvents with their properties are listed in Table 4.1.

Table 4.1 Common recrystallization solvents* and their properties

	Solvent	Boiling point (°C)	Properties
Polarity ↑	Water	100	Nonflammable, usually solvent of choice for many 'polar' compounds, Disadvantage-crystals dry slowly
	Methanol	64	Miscible with water, good for relatively polar compounds Advantage-easily removed
	95% Ethanol	78	Considered an excellent general solvent Preferred over methanol for higher b.p Disadvantage-contains 5% water
	Acetone	56	General purpose solvent for relatively polar compounds Disadvantage-low b.p. and highly inflammable- makes it difficult to work with
	Ethylacetate	77	Good general solvent for compounds with intermediate polarity
	Chloroform	61	Immiscible with water, excessive inhalation of vapours could be toxic
	Dichloromethane	41	General solvent for compounds with intermediate polarity Disadvantage-low bp, fairly hazardous
	Toluene	111	Good solvent for aromatic compounds Disadvantage-high b.p. makes it difficult to remove
	Petroleum ether	30–60	It is a mixture of hydrocarbons, good solvent for nonpolar compounds
	Cyclohexane	81	Good general solvent for nonpolar compounds
	Hexane	69	Good for nonpolar compounds, Advantage-easily removed

4.3.4 More on Solvents for Recrystallization

Other solvents used for recrystallization include ethyl methyl ketone (b.p. 80 °C); tetrahydrofuran (b.p. 66 °C); acetonitrile (b.p. 80 °C); chlorobenzene (b.p. 132 °C); diethyl ether (b.p. 36 °C) and carbon disulphide (b.p. 46 °C). However, the use of diethyl ether should be avoided as it is highly flammable and also its high volatility causes solid deposition by complete evaporation rather than preferential crystallization. Carbon disulphide has a low flash point and also forms explosive mixtures in air, thus its use should be completely avoided if an alternative solvent is available.

While choosing a suitable solvent for recrystallization, solubility of a small amount of the crude compound should be checked in the solvent. If the compound readily dissolves in the solvent at room temperature, the solvent cannot be used for recrystallization. If the compound is sparingly soluble or insoluble in the solvent at room temperature, the mixture should be heated to the boiling point of the solvent.

If the compound is still insoluble, it is considered insoluble at all temperatures and the solvent is not suitable for recrystallization and another solvent should be tested. However, if the compound completely dissolves in the solvent at or near its boiling point and crystallization happens on gradual cooling, the solvent can be employed for recrystallization of the crude compound. A two solvent or a mixed solvent recrystallization should be considered only when a single suitable solvent cannot be found for the purpose.

4.4 Pre-lab Questions

1. Why is there a need to recrystallize an organic compound?

   ....
   ....
   ....
   ....
   ....

2. Briefly explain the principle of recrystallization.

   ....
   ....
   ....
   ....
   ....

3. Out of the given statements, which is the correct criterion for selecting a solvent suitable for a single solvent recrystallization?
   The compound to be purified should be:
   a. soluble in hot solvent and soluble in cold solvent.
   b. insoluble in hot solvent and insoluble in cold solvent.
   c. insoluble in hot solvent and soluble in cold solvent.
   d. soluble in hot solvent and insoluble in cold solvent.

4. Explain what is meant by "like dissolves like".

   ....
   ....
   ....
   ....
   ....

5. What method can be adopted to remove coloured impurities from an organic compound to be purified?

....
....
....
....
....

6. Why is necessary to allow the hot solvent to cool gradually rather than immediate chilling during the process of recrystallization?

....
....
....
....
....

7. Make a table of all the organic compounds and solvents used during the recrystallization experiment including their chemical structure, handling precautions and health hazards.

....
....
....
....
....

4.5 Important Techniques to Be Learnt

4.5.1 Gravity Filtration

The most common type of filtration technique is gravity filtration which involves passing the solution to be filtered through a filter paper held in a funnel, allowing the gravity to draw the solution through the paper into a flask. It may be used for two purposes, either to collect the precipitated/recrystallized product from the solution or to remove solid impurities from a liquid.

Procedure: For gravity filtration, fold the filter paper into a cone as shown in Fig. 4.2. Place the filter paper into a funnel beneath which is kept the Erlenmeyer/ conical flask. To set the filter paper in place, wet it with a few drops of the solution to be filtered. Slowly, pour the solution to be filtered through the funnel collecting the filtrate in the conical flask and the product/ insoluble impurities on the filter paper.

4.5 Important Techniques to Be Learnt

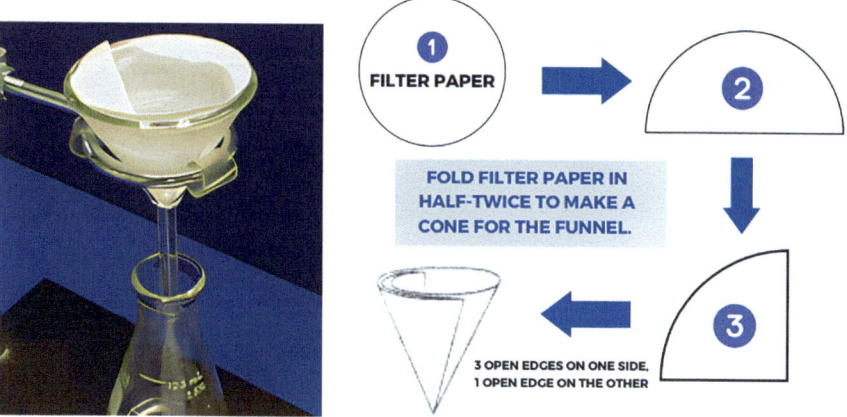

Fig. 4.2 Folding filter paper into a cone

4.5.2 Hot Filtration

Hot filtration is essentially used to separate insoluble impurities from a hot solution. It is a modified way of gravity filtration which is better carried out using a fluted filter paper and keeping the apparatus being used warm.

Procedure: To carry out hot filtration, preheat a short stemmed funnel and an Erlenmeyer flask in a hot air oven. A short stemmed funnel is used to avoid crystallization in the stem as the solution cools. Fold a fluted filter paper as shown in Fig. 4.3. Place the fluted filter paper in the warm funnel held over the Erlenmeyer flask. Fit the filter paper in the funnel by pouring some hot solvent. Now carefully pour the hot solution to be filtered through funnel. Some solvent may evaporate quickly depositing little amount of crystals on the filter paper along with the insoluble impurities. Wash down the filter paper with some hot solvent to redissolve these crystals, the impurities will remain undissolved. Allow the hot filtrate to gradually cool depositing the recrystallized product.

4.5.3 Vacuum/Suction Filtration

Filtration under suction or vacuum filtration is a fast and efficient way of filtration used to collect a product obtained from a synthesis or to collect recrystallized product from a solution. The solution to be filtered is drawn through a filter paper under vacuum rather than under gravity as described in previous methods. The process requires filtration unit (Fig. 4.4) comprising of Buchner funnel, filtration flask with a side outlet (for smaller quantities to be filtered, Buchner funnel is replaced by Hirsch

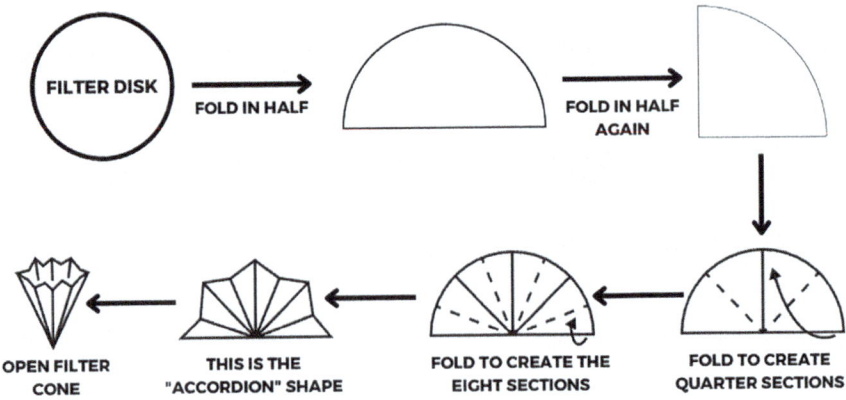

Fig. 4.3 Making a fluted filter paper

funnel and filtration tube with side outlet can replace filtration flask), conical filter adapter and a source of vacuum.

Procedure: Before beginning the filtration process check the filtration funnel and flask/ tube for any cracks, should there be any, the apparatus should be replaced. Clamp the filtration tube securely, place the conical filter adapter on the neck of the tube and fit in the Hirsch funnel. Place an appropriately cut filter paper in the Hirsch funnel so as to cover all the holes while not rising to the sides of the funnel. Wet the filter paper with some solvent or solution being used and connect the side outlet of the filter tube to the vacuum source. Now pour the solution to be filtered into the funnel taking care not to fill it to the brim. The solution will be drawn into the filtration tube quickly due to vacuum leaving the desired product/ crystals in the funnel. This may be washed with a small volume of cold solvent to remove any soluble impurities. Let the product dry under vacuum for some time. Disconnect the rubber tubing from

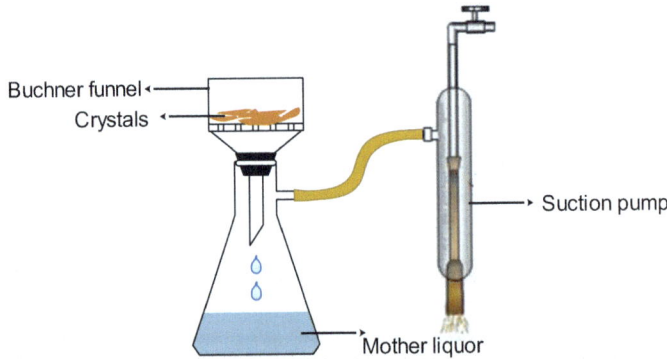

Fig. 4.4 Vacuum filtration unit

the filter tube outlet before turning off the vacuum source. Lift the filter paper along with the product from the funnel and scrape off the product.

4.6 Experimental Section

4.6.1 Chemicals and Apparatus/Equipment

1. Benzoic acid, naphthalene, *m*-dinitrobenzene, acetanilide.
2. Distilled water, ethanol.
3. 250 mL Erlenmeyer flask, short stemmed funnel, Hirsch funnel, filter tube, spatula filter paper, thermometer, melting point tubes, porous plate, spatula, buret drop tube, mortar and pestle boiling stones, suction pump.

4.6.2 Experimental Procedure

4.6.2.1 Single Solvent Recrystallization

Recrystallization of benzoic acid with water as a solvent

Take the given amount of benzoic acid and water in a conical flask or a boiling tube. Heat the contents on wire gauze using a Bunsen burner. Continue to heat till the benzoic acid is completely dissolved giving a clear solution. In case the solution is not colourless, allow it cool slightly, add a small amount (0.25 g) of activated charcoal and continue to boil for a few minutes so as to remove any coloured impurities. Filter the hot solution through a fluted filter paper (c/o hot filtration Sect. 4.5.2) placed in a short stemmed funnel and collect the filtrate in a conical flask. Allow the filtrate to cool gradually at room temperature so as to obtain the crystals of benzoic acid. Filter the crystals so obtained at suction using a Hirsch funnel as described under suction filtration. Scrape off the crystals and allow them to dry in air. The purity of the crystals can be checked by determining the melting point and comparing it with the literature value.

Recrystallization of *m*-dinitrobenzene from alcohol as a solvent

Take the given amount of *m*-dinitrobenzene (yellow solid) and ethanol in a boiling tube. Heat the contents in a water bath over Bunsen burner to obtain a clear solution. Filter the hot solution through a fluted filter paper (c/o hot filtration) placed in a short stemmed funnel and collect the filtrate in a conical flask. Allow the filtrate to cool gradually at room temperature so as to obtain the crystals of *m*-dinitrobenzene. Filter the crystals so obtained at suction using a Hirsch funnel as described under suction filtration. Scrape off the crystals and allow them to dry in air. The purity of the crystals can be checked by determining the melting point and comparing it with the literature value.

4.6.2.2 Mixed Solvent Recrystallization

If a single solvent is not able to meet the desired solubility characteristics of a compound, it is then advisable to make use of a mixed solvent system for recrystallization. The two solvents chosen for the system should be miscible in one another. Some common solvent pairs that can be employed for recrystallization are: ethanol–water, methanol–water, acetone–water, acetic acid–water etc. The compound to be recrystallized is first dissolved in minimum amount of the solvent in which it is soluble by heating to its boiling point followed by addition of the second solvent in which the compound is relatively insoluble. The second solvent is added drop-wise until the solution just becomes cloudy or turbid indicating precipitation. To bring the solution to saturation, a small amount of the first solvent is added to barely obtain a clear solution. At this point, the solution is saturated and crystals should start separating on gradual cooling. The solution should not be cooled rapidly and care should be taken not to add excess of the second solvent that might result in separation of the compound as oil or a viscous liquid. If it so happens, the solution should be heated again by adding small amount of the first solvent.

Recrystallization of acetanilide using alcohol and water as mixed solvents

Take the given amount of acetanilide and dissolve it in small amount of ethanol by heating in a water bath. In case some insoluble impurities are present filter the solution. To the filtrate add warm water drop-wise just enough to bring slight turbidity/cloudiness. Clear the turbidity by adding minimum amount of warm ethanol drop-wise. Leave the clear solution undisturbed to gradually cool at room temperature so as to obtain the crystals of acetanilide. Filter the crystals so obtained at suction using a Hirsch funnel and wash them with ice-cold water. Scrape off the crystals and allow them to dry in air. The purity of the crystals can be checked by determining the melting point and comparing it with the literature value.

4.7 Post-lab Questions

1. The solubility of compound "X" in ethanol is 0.80 g per 100 ml at 0 °C and 5.0 g per 100 mL at 78°C. What is the minimum amount of ethanol needed to recrystallize a 12.00 g sample of compound "X"? How much would be lost in the recrystallization, that is, would remain in the cold solvent?

   ....
   ....
   ....
   ....
   ....
   ....

2. While removing coloured impurities during recrystallization, why is it necessary to cool the solution before adding activated charcoal?

   ....
   ....
   ....
   ....
   ....
   ....

3. Why Erlenmeyer flask or boiling tube is chosen over beaker to carry out recrystallization?

   ....
   ....
   ....
   ....
   ....
   ....

4. Why a short stemmed funnel is used while carrying out hot filtration?

   ....
   ....
   ....
   ....
   ....
   ....

5. Compound A is highly soluble in toluene, but only slightly soluble in petroleum ether. How can these two solvents be used in combination in order to recrystallize Compound A?

....
....
....
....
....
....

6. Would addition of excess activated charcoal for decolourization have any effect on the yield of the recrystallized product?

....
....
....
....
....
....

7. You have prepared a compound B which is reported to have a pale yellow color. The hot solution while carrying out recrystallization of B becomes yellow. Should you use decolorizing charcoal before allowing the hot solution to cool? Explain your answer.

....
....
....
....
....
....

4.7 Post-lab Questions

Flowchart for Recrystallization

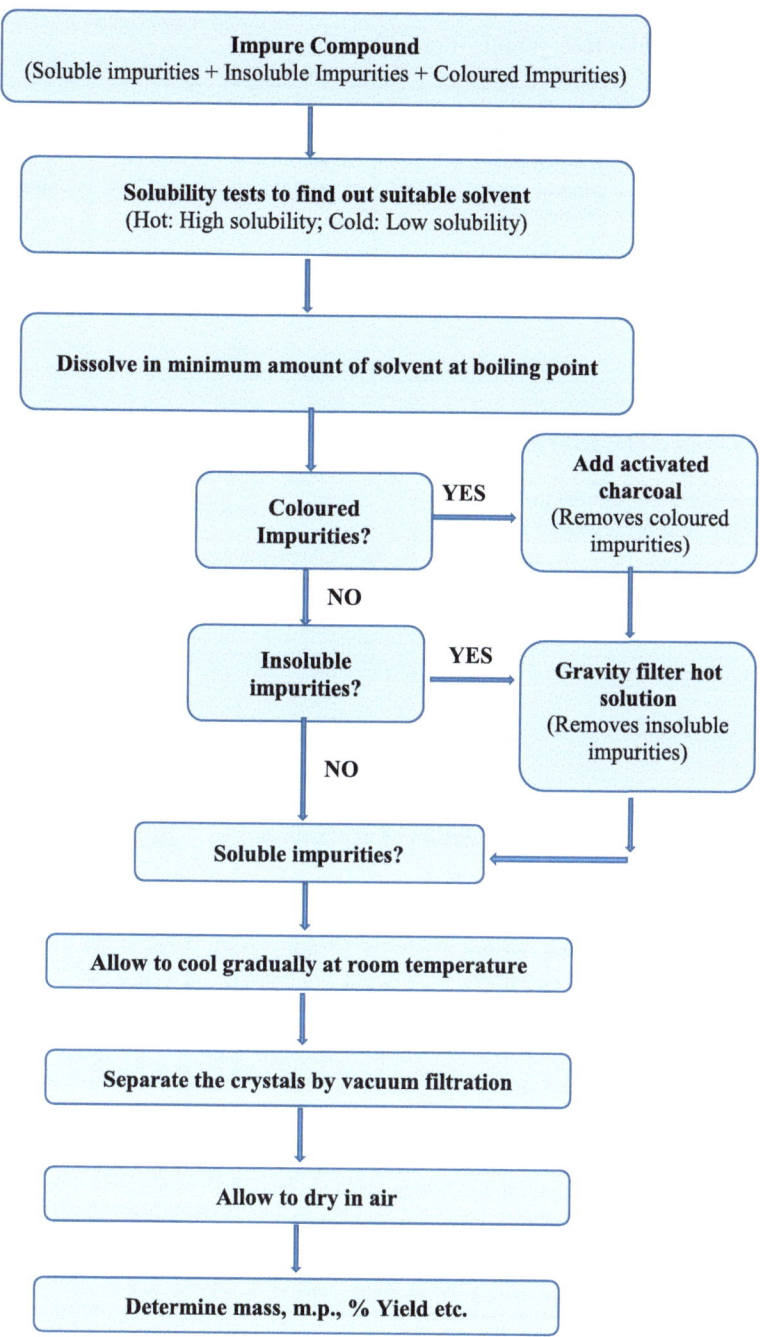

4.8 Summary Sheet

Observation table: Recrystallization of benzoic acid

Recrystallization solvent used					
Time allowed for crystals to form					
Mass of impure benzoic acid (g)	Mass of pure benzoic acid recovered (g)	% Yield	Appearance of crystals	Observed m.p. °C	Literature m.p. °C

Observation table: Recrystallization of *m*-dinitrobenzene

Recrystallization solvent used					
Time allowed for crystals to form					
Mass of impure *m*-dinitrobenzene (g)	Mass of pure *m*-dinitrobenzene recovered (g)	% Yield	Appearance of crystals	Observed m.p. °C	Literature m.p. °C

Observation table: Recrystallization of acetanilide

Recrystallization solvent used					
Time allowed for crystals to form					
Mass of impure acetanilide (g)	Mass of pure acetanilide recovered (g)	% Yield	Appearance of crystals	Observed m.p. °C	Literature m.p. °C

Results and Discussion

....
....

4.8 Summary Sheet

Notes

Chapter 5
Determination of Boiling Point of Liquid Compounds

5.1 Objectives

To determine the boiling point of the given organic liquid.
To learn the distillation, capillary and BODMEL's methods for determination of boiling points.

5.2 About the Experiment

Through this experiment you would learn the technique of simple distillation for purification of organic liquids and also finding out their boiling points. Siwoloboff's method (capillary method) as well as the BODMEL's method would also be illustrated as a simplified set up for boiling point determination.

5.3 Introduction

5.3.1 Basics of Distillation

What recrystallization is to organic solids, distillation is to organic liquids. In the previous experiments you had learnt that a solid organic compound can be purified by recrystallization and identified by melting point determination. Similarly, purification and identification of an organic liquid compound can be done by distillation which in itself is a way of determining the boiling point.

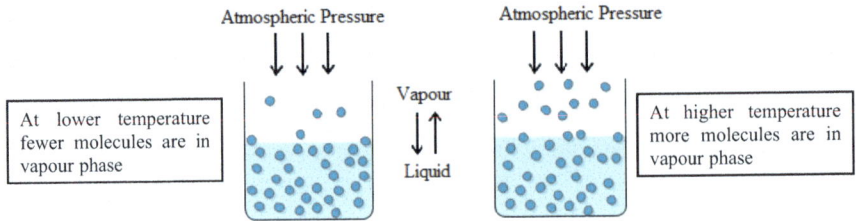

Fig. 5.1 Effect of temperature on liquid–vapour equilibrium

To understand the process of distillation, let's consider what happens to a liquid upon heating. At any given temperature, there exists an equilibrium between the molecules of a liquid escaping into the vapour phase and the ones condensing back to the liquid phase (Fig. 5.1). As the temperature increases more molecules possess higher kinetic energy to escape, resulting in a greater number of molecules being present in the vapour phase.

One can have measure of this degree of vaporization if the liquid is placed in a closed container. At a given temperature, this equilibrium pressure is defined as the vapour pressure of the compound and as the temperature is increased, the vapour pressure of the liquid also increases. At a particular temperature, when the vapour pressure (escaping tendency) of the liquid becomes equal to the atmospheric pressure (1 atm = 760 mm Hg), it then begins to boil. This temperature is then recorded as the boiling point of the liquid and is termed as *normal boiling point*. For example, the normal boiling point for toluene is 110 °C (Fig. 5.2). When a pure liquid boils; the liquid is converted to vapour rapidly. Even if the rate of heating increases, the temperature of the boiling liquid will not change, only the rate of vaporization may increase as the energy supplied by heating is used to cause the liquid–vapour phase change.

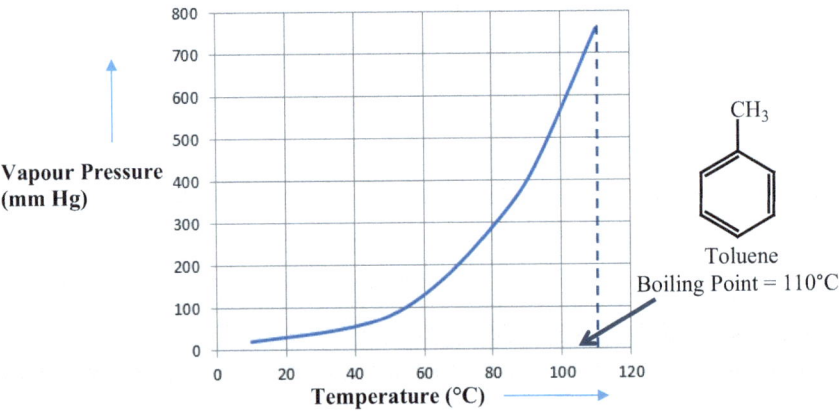

Fig. 5.2 Temperature dependence of vapour pressure for toluene

5.3 Introduction

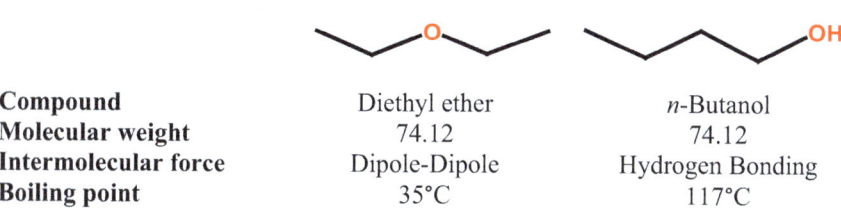

Compound	Diethyl ether	n-Butanol
Molecular weight	74.12	74.12
Intermolecular force	Dipole-Dipole	Hydrogen Bonding
Boiling point	35°C	117°C

Fig. 5.3 Effect of intermolecular forces on boiling point

5.3.2 Factors that Affect Boiling Point of Liquids

It may be noted that the boiling point of a liquid is pressure dependent. Under vacuum i.e. at reduced pressure, the boiling point for a liquid will be lower than the boiling point at atmospheric pressure. Similarly, increasing the pressure will increase the boiling point of a liquid.

Apart from external pressure, there are different other factors that affect the boiling point of liquid compounds:

5.3.2.1 Intermolecular Forces of Attraction

The relative strength of the intermolecular forces, ionic > hydrogen bonding > dipole-dipole > Van der Waals dispersion forces, is fundamental to explaining as to why different liquids have different boiling points. Consider the two compounds, diethyl ether and n-butanol that have the same molecular weight (Fig. 5.3). The boiling point of diethyl ether is 35 °C and its molecules are held together by dipole–dipole interactions which arise due to the polarized C-O bonds. Whereas, its isomer n-butanol has a greatly increased boiling point of 117 °C due to the fact that butanol contains a hydroxyl group, which is capable of hydrogen bonding. Thus, depending upon the functional groups present, the intermolecular forces of attraction largely influence the boiling points of liquids.

5.3.2.2 Molecular Weight

For molecules with the same functional group, the boiling point increases with increase in the molecular weight. This can be explained on the basis of Van der Waals dispersion forces, which are proportional to surface area. An increase in the length of the chain will also increase the surface area meaning thereby that the ability of individual molecules to attract each other increases. Thus higher temperature is required to pull apart the molecules of the compound transforming it from liquid to the gas phase, leading to increase in the boiling point. For example, the series of alcohols with increasing chain length have increasing boiling points as shown in Fig. 5.4.

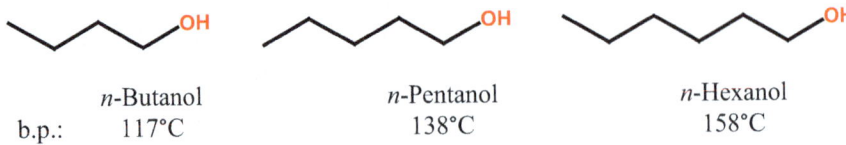

	n-Butanol	*n*-Pentanol	*n*-Hexanol
b.p.:	117°C	138°C	158°C

Fig. 5.4 Effect of molecular weight on boiling point

Fig. 5.5 Effect of branching on boiling point

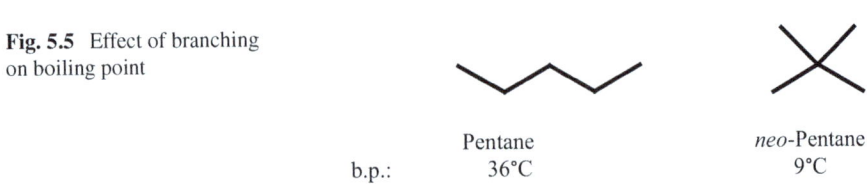

	Pentane	*neo*-Pentane
b.p.:	36°C	9°C

5.3.2.3 Branching

As branching increases, the surface area decreases and hence fewer Van der Waals interactions operate between molecules. As a result of weakened intermolecular forces of attraction, the boiling point decreases. Pentane, a linear molecule has a higher boiling point of 36 °C compared to its branched isomer, *neo*-pentane which has a boiling point of 9 °C (Fig. 5.5).

Furthermore, the boiling range for a pure liquid compound will be sharp and narrow, while the boiling range for an impure liquid compound will depend upon the nature and quantities of impurities present. Insoluble/ non-volatile impurities like sand will have no effect on the boiling point and such solid impurities are left behind after the liquid has evaporated. However presence of volatile impurities usually elevates the boiling point or in case the liquid mixture forms an azeotrope, the boiling point remains constant. Such azeotropic mixtures, mimic the boiling behavior of pure liquids and distill at a constant boiling temperature. For example a 96% ethanol-4% water mixture is a constant boiling azeotrope at 78.1 °C.

5.3.3 Distillation: Uses and Types

The basic process of distillation involves heating a liquid to its boiling point at atmospheric or reduced pressure such that the liquid molecules vapourize. The vapours on passing through a condenser get cooled and return to the liquid phase in the pure form that can then be collected. The technique of distillation is commonly employed for:

1. Purification of liquids.
2. Determining the boiling points of liquid compounds and hence for identification of unknowns.

3. Separating mixtures of liquids into their individual components based on the difference in volatilities of the individual components.

Commercial applications of distillation include refining of crude oil into its fractions, manufacture and purification of nitrogen, oxygen and rare gases, desalination of water etc. Depending upon its applicability in purification and separation of liquids, the methods of distillation can be classified into three basic types:

5.3.3.1 Simple Distillation

It is used to separate volatile solvent/compound from non-volatile compound (product/impurity) or to separate two liquids differing in their boiling points by at least 25 °C. The technique is also employed for determining the boiling point of a liquid.

5.3.3.2 Fractional Distillation

It is useful in separation of liquid compounds having their boiling points less than 25 °C apart.

5.3.3.3 Steam and Vacuum Distillations

These are used for purification and or separation of high boiling liquids. Vacuum distillation or distillation under reduced pressure is especially useful for the compounds that tend to decompose at or near their normal boiling point.

5.4 Pre-lab Questions

1. Define normal boiling point of a liquid.
   ....
   ....
   ....
   ....

2. The normal boiling point of cyclohexane is 81°C. What will be the vapor pressure of cyclohexane at 81°C?
   ....
   ....
   ....
   ....

3. For a particular sample, will the boiling point be unchanged, increase, or decrease if you try to distill it under a reduced pressure?
   ....
   ....
   ....
   ....

4. Briefly describe the different factors that affect the boiling point of a compound.
   ....
   ....
   ....
   ....

5. What will be the effect on the boiling point of water if sand is present as an impurity?
   ....
   ....
   ....
   ....

6. You need to separate a mixture of two liquids X and Y having boiling points 122°C and 135°C respectively. What method will you adopt?
   ....
   ....
   ....
   ....

7. If a liquid sample distils at a constant boiling temperature, is it certainly pure?
   ....
   ....
   ....
   ....

5.5 Experimental Section

5.5.1 *Chemicals and Apparatus/Equipment Required*

Chemicals	Apparatus/Equipment
Liquid compound A (boiling range below 100 °C) Liquid compound B (boiling range above 100 °C)	Distillation method: Round bottom flask, still head, mercury pocket, thermometer, Liebig condenser, receiver adapter, receiving flask, sand bath, water bath, tripod stand, clamp stands, Bunsen burner, rubber tubing Siwoloboff's method: Kjeldahl's flask with bath liquid, thermometer, ignition tubes, capillary tubes, Bunsen burner, sand bath, dropper, rubber bands BODMEL's method:BODMEL'S Apparatus, glass dropper, boiling chips

5.5.2 *Experimental Procedure*

In this experiment you will be given two liquid compounds, Liquid A having a boiling range below 100 °C and Liquid B having a boiling range above 100 °C. You will have to determine the boiling range for each of the two compounds by distillation, Siwoloboff's method (capillary method) and BODMEL's Method.

5.5.2.1 Boiling Point Determination by Distillation Method

To determine boiling point of liquid compounds, simple distillation method is usually employed when sufficient amount of the sample is available. The quick fit apparatus used for the purpose is shown in Fig. 5.6.

To set up the distillation assembly, pour the given liquid compound A (boiling range below 100 °C) into the round bottomed flask along with a few pumice stones. The flask should be filled not more than half. Clamp the flask over a water bath placed on a tripod stand above the Bunsen burner. Fit the still head into the neck of the flask and attach the Liebig condenser to it. To the other end of the condenser attach the receiver adapter and place the receiving flask to collect the distillate below it. Insert the mercury pocket along with the thermometer into the still head and attach the tubing to water inlet and outlet of the condenser. Inspect the complete assembly to ensure that no joint is under stress and the system can be safely heated. Turn on the water supply to maintain a steady flow of water through the condenser. Begin to heat the liquid by turning on the burner; the liquid will eventually begin to boil. Boiling

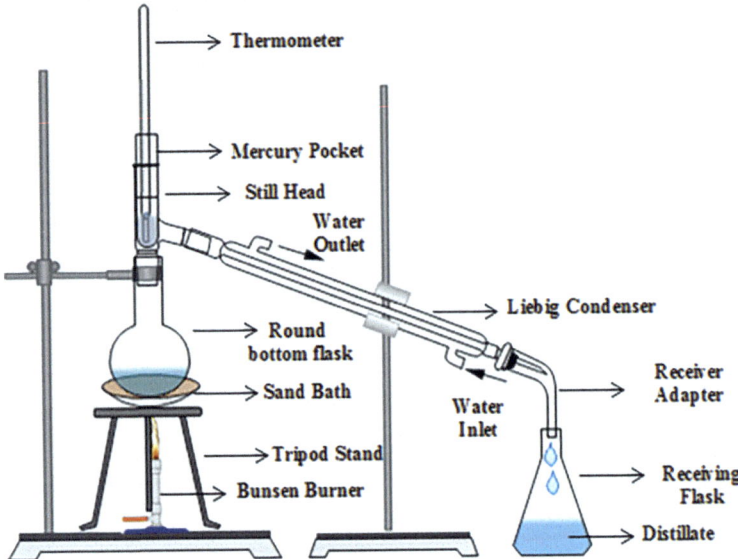

Fig. 5.6 Apparatus assembly for simple distillation

should be gentle enough so that the hot vapors move slowly up into the distilling head reaching the thermometer bulb. Soon the liquid will begin to condense in the side-arm of the distilling head and thereafter drip into the receiving flask. Most of the liquid would distill off at a constant temperature range. Record this temperature as the boiling point of the liquid. Stop heating when little liquid remains in the distilling flask and raise the apparatus out of the water bath. Allow it to cool before disassembling it.

To determine the boiling point of Liquid compound B (boiling range below 100 °C), perform the same procedure as above by taking Liquid B in the round bottom/ distilling flask and replacing the water bath by a sand bath.

Some important distillation tips:

- Remember to inspect all glassware especially the distillation round bottom flask for any star-cracks.
- Always use pumice stones/ boiling stones/ boiling chips while carrying out distillation. It prevents bumping of the liquid and ensures smooth boiling. These boiling stones are porous and contain many air filled pores. As the vapours of the liquid getting distilled penetrate the pores, air is gradually pushed out from them. This steady escape of air from the boiling stones results in bringing a smooth and uniform boil to the liquid.
- Never add pumice stones to an already hot solution. Doing so may result in a violent degassing of the liquid making it splash out of the set up.

- *Never distill a solution to dryness trying to get the last drop. The distillation flask will begin to superheat creating flammable conditions for the vapors in the apparatus.*

5.5.2.2 Boiling Point Determination by Siwoloboff's Method (Capillary Method)

Siwoloboff's method of determining the boiling point of liquid compounds is employed when the amount of the liquid sample available is small. The apparatus set up used for the purpose is shown in Fig. 5.7.

Take about 0.5 mL of the given liquid compound A in an ignition tube. Introduce an inverted capillary with its open end into the liquid as shown in Fig. 5.7 (inset a). Securely attach this ignition tube to the lower end of thermometer with the help of a rubber band and insert it into the Kjeldahl's flask containing the bath liquid. The rubber band should be well above the level of the bath liquid. Heat the set up with the help of a Bunsen burner uniformly. After a while slow and erratic escape of bubbles from the open end of the capillary will be observed (Fig. 5.7 inset b). Soon after that a rapid and continuous stream of bubbles can be seen escaping from the capillary and liquid begins to rise in the capillary (Fig. 5.7 inset c). Record the temperature observed at this point of time as the boiling point. Allow the apparatus to cool to room temperature and remove the ignition tube from the thermometer. Repeat the same procedure for liquid compound B using fresh capillary and ignition tubes.

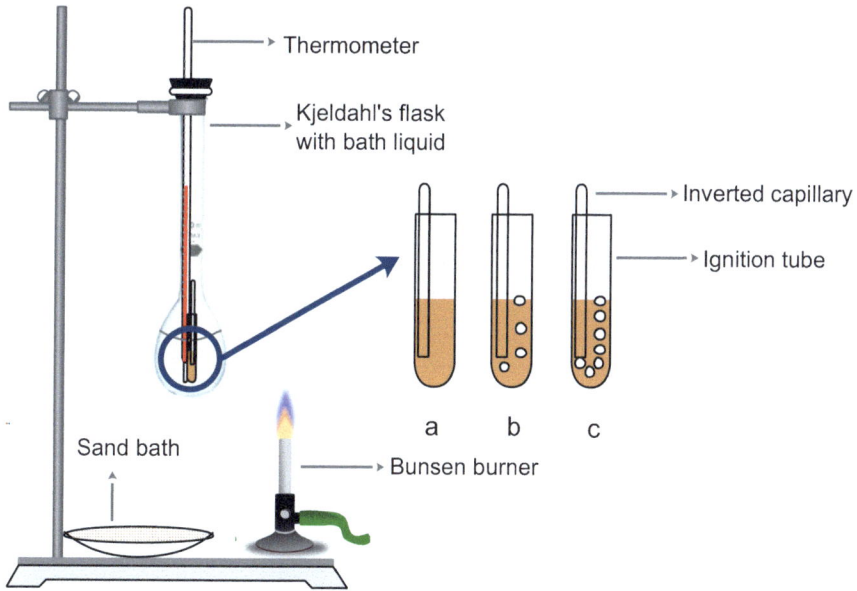

Fig. 5.7 Boiling point determination by Siwoloboff's method

5.5.2.3 Boiling Point Determination by BODMEL's Method

BODMEL's method, involving the use of a safer and handy set up of a new single piece of glass apparatus is employed to acquire the benefits of reduction in scale at affordable cost in the teaching laboratories.

For the determination of boiling point, the apparatus needs to be clamped securely in such a manner that the finger like projections rest slightly above the clamp as evident from Fig. 2.2. Introduce the given liquid (1–2 mL) of which boiling point is to be determined in the inner tube with the help of a glass dropper taking precautions that the liquid does not enter into the side projections. In order to ensure smooth boiling, put a tiny piece of boiling chip or pumic stone and thereafter introduce the cork carrying thermometer ensuring that the bulb of thermometer rests in the constriction part of the apparatus. Thereafter, with the help of low flame Bunsen burner, heat the outer jacket of BODMEL apparatus uniformly. Once the boiling point is achieved, the temperature becomes constant and the liquid starts distilling and getting collected in the projections. The liquid can be collected with the aid of a dropper.

5.6 Post-lab Questions

1. How does adding boiling chips aid distillation?

   ....
   ....
   ....
   ....
   ....

2. What is the purpose of using a condenser during the process of distillation?

   ....
   ....
   ....
   ....
   ....

3. You suddenly notice you have forgotten to add boiling stones to your distillation flask, but the distillation is now in progress. What should you do?

   ....
   ....
   ....
   ....
   ....

5.6 Post-lab Questions

4. Under what conditions is it advisable to use the Siwoloboff's method for boiling point determination?

   ....
   ....
   ....
   ....
   ....

5. Why should the distillation flask be never heated to dryness?

   ....
   ....
   ....
   ....
   ....

6. While carrying out distillation when should one employ a water bath and when should a sand bath be used?

   ....
   ....
   ....
   ....
   ....

7. Explain in brief the basic principle behind the process of distillation.

   ....
   ....
   ....
   ....
   ....

8. What are the advantages of using BODMEL's apparatus for determining boiling point of liquids?

   ....
   ....
   ....
   ....
   ....

5.7 Summary Sheet

A. Boiling point determination by distillation method

Sample	Observed boiling point (°C)	Literature boiling point (°C)
Liquid A		
Liquid B		

B. Boiling Point Determination by Siwoloboff's Method

Sample	Observed boiling point (°C)	Literature boiling point (°C)
Liquid A		
Liquid B		

C. Boiling Point Determination by BODMEL's Method

Sample	Observed boiling point (°C)	Literature boiling point (°C)
Liquid A		
Liquid B		

Results and Discussion

....

....

Notes

Chapter 6
Chromatography

6.1 Objectives

To learn the technique of chromatography for separation of organic compounds.
To separate the given mixtures of compounds by paper chromatography and thin layer chromatography (TLC).

6.2 About the Experiment

This experiment would make you adept with the technique of chromatography for separation of organic compounds. Principles and methods of separation of mixture of organic compounds like amino acids, carbohydrates and nitro/ amino phenols by paper chromatography (ascending & horizontal/ radial) and thin layer chromatography have been illustrated.

6.3 Introduction

"Like light rays in the spectrum, the different components of a pigment mixture, obeying a law, are resolved on the calcium carbonate column and then can be qualitatively and quantitatively determined. I call such a preparation a chromatogram and the corresponding method the chromatographic method."

—M. S. Tswett

https://en.wikipedia.org/wiki/Mikhail_Tsvet#/media/File:Mikhail_Tsvet.jpg

Originating from two Greek roots, *chroma* (colour) and *graphé* (writing), the verbatim translation of the word chromatography means 'colour writing'. Russian botanist, Mikhail Semenovich Tswett invented the technique and coined the term chromatography while working on separation of coloured components of the green plant pigment chlorophyll.

By definition, chromatography is a technique used for separation of a mixture by passing it in a solution form through an adsorbent medium in which its components move at different rates. Various types of chromatographic techniques have been developed and extensively used by scientists for separation, purification, identification and characterization of individual components of mixtures that may or may not be coloured. One of the major advantages of chromatography over other purification and separation processes is that it is a particularly useful method when separation is to be carried out on a micro-scale, i.e. when components are available in minute amounts.

6.3.1 *Chromatographic Techniques: Terminology and Classification*

Before delving deep into the principles and technique of chromatography, let us understand a few important terms that will keep coming across while going through the topic.

Adsorbent: An adsorbent is the first thing required to carry out separation using chromatography. It is essentially a porous material that can suck up liquids and solutions so as to form a surface layer. Paper, silica gel, alumina (ultrafine aluminum oxide) are a few adsorbents routinely used for chromatographic separations.

Solid support: An adsorbent can be a self-supporting one or may need a solid support. Paper is stiff and can stand up by itself however; adsorbents like silica and alumina are powders and need a solid support to hold them. A glass plate, aluminium or plastic sheets of suitable size can serve as solid support for such adsorbents which when coated as a thin layer on the support surface form the basis of thin layer chromatography (TLC).

Stationary phase: The adsorbent and solid support (wherever required) together make the stationary phase of a chromatographic system on which the components to be separated are selectively adsorbed.

Mobile phase: It is the liquid or gas that while flowing through the chromatographic system carries along the components to be separated at different rates over the stationary phase.

Sample or analyte: The substance of interest that is being analysed for separation into its individual components is termed as sample or analyte. It is also sometimes referred to as solute.

Eluent: To carry out the separation of the analyte, certain solvents like petroleum ether, acetone, ethylacetate etc. are required which in chromatographic terminology are called eluents.

Elutropic series: Since the components of the analyte have different degrees of affinity for the adsorbent, the polarity of the solvent/ eluent plays an important role in separation. Thus elutropic series is nothing but a list of such common eluents arranged in increasing order of their polarity/ eluting power. Enlisted below are some solvents arranged according to increasing polarity:

Petroleum ether - Dichloromethane - Ethylacetate - Acetone - Methanol

→ Increasing Polarity

Figure 6.1 illustrates the simple chromatographic separation of black ink into its constituent blue and red pigments by using paper as the stationary phase and water as the mobile phase.

Having understood some common chromatographic terms, it will now be easier to go through a little more technical detail. Chromatographic separation is principled upon the differential distribution of the components of a mixture between two immiscible substances one being a mobile phase and the other a stationary phase. In other words, all the chromatographic techniques depend on the differential partition of solutes between these two distinctive phases. Such kind of partition is described

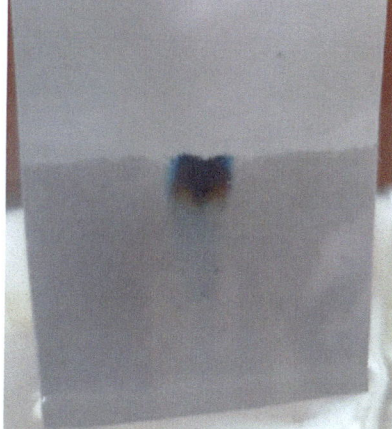

Fig. 6.1 Black ink is not that black!

by "distribution coefficient" or "partition co-efficient" which is constant at a given temperature for the two phases.

To understand this principle, let us suppose we have an analyte "A" which is transferred between the mobile and stationary phases.

$$A_{mobile} \longleftrightarrow A_{stationary}$$

The equilibrium constant for this reaction is given by "K_D" known as the distribution constant or partition coefficient.

$K_D = [A]_s/[A]_m$ where,

$[A]_s$ = molar concentration of analyte A in stationary phase

$[A]_m$ = molar concentration of analyte A in mobile phase

The concentration of the components could be expressed in terms of either mass per unit volume or mass per unit mass depending on the technique employed. It is important to understand that during the process of chromatographic separation, the analytes incessantly keep moving back and forth between the two phases so that differences in their distribution coefficients would ultimately result in their separation.

The mobile phase that may be a liquid or a gas carries along the components of the mixture while passing through the stationary phase that may be a solid or a liquid.

While the underlying principle is decided by the physical nature of these two phases; there can be various ways of classifying the chromatographic methods. Table 6.1 gives classification of some common chromatographic methods based on the physical means of bringing the stationary and the mobile phases into contact along with the principle involved.

Adsorption chromatography involves a *solid–liquid* combination wherein the solid stationary phase tends to adsorb the components of the mixture under the influence of the liquid mobile phase and separation is achieved depending upon

Table 6.1 Common chromatographic methods

General classification	Stationary phase	Mobile phase	Principle involved
Column chromatography	The stationary phase is held in a narrow tube through which the mobile phase is forced either by pressure or by gravity.		
• Simple column	Solid	Liquid	Adsorption
• HPLC	Solid	Liquid	Adsorption
• GLC	Liquid	Gas	Partition
Planar chromatography	The stationary phase is supported on a flat plate or in the fibres of a paper. Here the mobile phase moves through the stationary phase by capillary action or by gravity.		
• Paper chromatography	Liquid	Liquid	Partition
• TLC	Solid	Liquid	Adsorption

HPLC: High Performance Liquid Chromatography; GLC: Gas Liquid Chromatography; TLC: Thin Layer Chromatography

6.3 Introduction

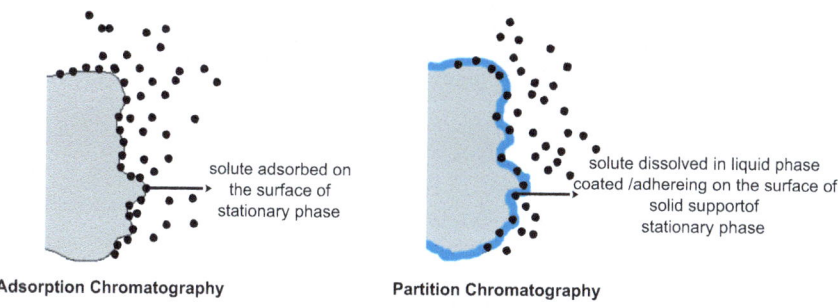

Fig. 6.2 Difference between adsorption and partition chromatography

the relative adsorption propensity of the components. Thin Layer Chromatography (TLC) is one such separation technique that is based on the principle of adsorption. Whereas, paper chromatography is an example of partition chromatography wherein, separation depends upon the differential partitioning of the components of the mixture between the stationary and the mobile phases that are both liquids. *Partition chromatography* may also involve a combination of liquid stationary phase against a gas mobile phase that forms the basis of GLC (Gas Liquid Chromatography). Figure 6.2 gives a pictorial representation of the principle difference between adsorption and partition chromatography. Apart from the common chromatographic methods mentioned in Table 6.1, ion-exchange chromatography, molecular exclusion chromatography and affinity chromatography etc. are also used as separation techniques. However, these are beyond the scope of this unit and we will be focusing on TLC and paper chromatography in detail.

6.3.2 Paper Chromatography

This type of planar chromatography works on the principle of differential partition of solutes between two liquid phases. For paper chromatography one of these two liquid phases is the water bound to the cellulose of paper and hence serves as the stationary liquid phase. The other liquid phase is the solvent of choice that flows along the length of paper by capillary action to serve as the mobile phase. In ascending paper chromatography, the solute mixture to be separated is spotted at a base line on the Whattman filter paper (No. 1) and the mobile phase is allowed to run/ flow upwards along the length of the paper through the spotted mixture by capillary action. While in descending paper chromatography the mobile phase passes through the spotted mixture downwards by the action of gravity. As the mobile phase (solvent) flows through the solute mixture, the individual solutes tend to partition themselves between the aqueous stationary phase and the organic mobile phase depending upon their respective solubilities in the two phases. A solute more soluble in the organic mobile phase travels faster along the length of the paper leaving the lesser soluble

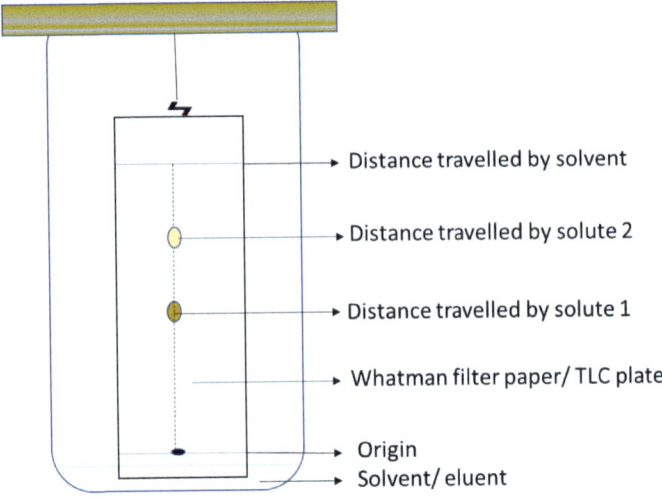

Fig. 6.3 Separation of solutes by paper/thin layer chromatography

solute behind, thereby bringing separation (Fig. 6.3). The individual spots of the separated components may be visible to the naked eye if they are coloured or can be developed by different techniques like spraying with a visualizing agent. The separation so achieved is measured in terms of retention factor R_f, that denotes the relative rate of flow/retention of a solute with respect to the solvent. The R_f value for a compound remains constant in a given solvent system if the experimental conditions remain unaltered.

$$\text{Retention factor, } R_f = \frac{\text{Distance travelled by solute}}{\text{Distance travelled by solvent}}$$

6.3.3 Thin Layer Chromatography (TLC)

Thin layer chromatography follows the principle of adsorption. Here the stationary phase is silica/alumina gel applied on a solid support of thin sheet of plastic/glass or aluminum which is called TLC plate. The mobile phase is a solvent of choice that travels along the TLC plate by capillary action. The silica/alumina gel applied on the TLC plate constitutes of a three-dimensional framework of alternating silicon/aluminium and oxygen bonds along with hydroxy bonds on the surface (Fig. 6.5). This network makes the coating highly polar and capable of making hydrogen bonds. As the solvent passes through the solute mixture spotted at a baseline marked on the TLC, an equilibrium is set up as silica tends to hold the solute by adsorption and the solvent tends to dissolve and move the solute along with it. Polarity of solvent, TLC

6.4 Pre-lab Questions

Fig. 6.4 UV Chamber for viewing TLC

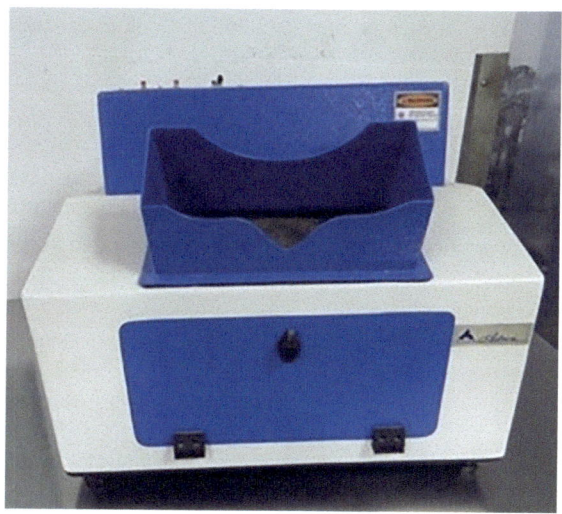

Fig. 6.5 Three-dimensional network of silicon and oxygen atoms in silica gel

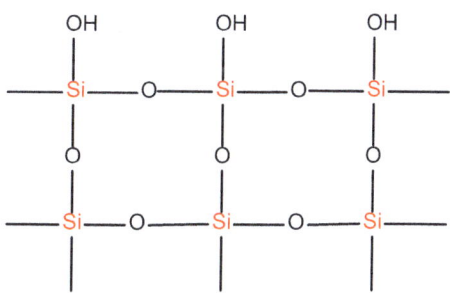

plate and strength of intermolecular forces of spotted solute influence the state of equilibrium and decide the ability of solvent to take the solute along as it travels up the plate. A more polar component tends to remain adsorbed on silica and therefore travels a shorter distance as compared to a solute that is less polar, thereby resulting in a separation (Fig. 6.3). The separated components can be viewed by placing the TLC in UV chamber (Fig. 6.4) or by placing it in a jar containing iodine crystals.

6.4 Pre-lab Questions

1. Which of the two compounds will have a higher value of R_f for a chromatography system with a polar stationary phase and a non-polar mobile phase?

2. On what factors does Rf value of a solute depend?

   ....
   ....
   ....
   ....

3. Why should a chromatography paper/ plate be marked with a pencil and not a pen?

   ....
   ....
   ....

4. For chromatographic separation of a mixture of two compounds, the following data is available. Distance travelled by solvent front is 5 cm from base line, while compounds A and B travel 2.5 cm and 4 cm respectively. What are the R_f values for compounds A and B?

   ....
   ....
   ....
   ....
   ....

5. Mark the following statements as true or false:
   a. A polar compound binds tightly to a polar stationary phase.
   b. The R_f value for a compound does not depend on the solvent chosen to run the chromatogram.
   c. Paper chromatography has a liquid stationary phase.

6.5 Experimental Details

6.5.1 Separation of Mixture of Sugars by Ascending Paper Chromatography

Here the separation of sugars will be achieved due to their partitioning between liquid stationary phase (water bound to cellulose of filter paper) and the liquid mobile phase (developing agent). The sugar structures contain several hydroxy bonds that are capable of hydrogen bonding with aqueous medium of the stationary or mobile phases. Their partition coefficients thereby favour aqueous phase and the R_f values for sugars become largely dependent on the nature of developers used. With non-aqueous developers, sugars tend to travel a shorter distance along the chromatogram hence displaying smaller R_f values, while developers having higher aqueous ratio result in larger R_f values for sugars. Further, the R_f values will also depend upon number of hydroxyl groups present in the structure, molecular mass and other functional groups that may be present in the structure of sugars.

6.5.1.1 Chemicals and Apparatus/Equipment Required

Chemicals	Apparatus/Equipment
Aqueous solutions of D-xylose, D-fructose and mixture of the two (Any other combination of sugars may be used viz *a* viz, D-glucose, D-mannose, D-galactose, maltose, sucrose, lactose etc.) Developing agent: Ethylacetate, acetic acid and water (54:25:12) Visualizing agent: Aniline hydrogen phthalate *(prepared by adding aniline (10 mL) and phthalic acid (1 g) to water-saturated butanol (40 mL, and water (60 mL)*	Whatman filter paper no. 1, spotting capillary tubes, chromatography jar with lid and holder

6.5.1.2 Experimental Procedure

Add appropriate volume of the developing agent (mobile phase) into the bottom of chromatography jar and cover with a lid to allow the chamber to get saturated with the solvent. Cut the Whatman paper intro chromatogram strips of required size. With the help of a pencil, draw a fine horizontal line about 1 cm away from the lower edge of the paper across its width. Mark three points on this base line and label them A, B and C. With the help of different spotting capillaries for each solution, apply the solution of D-xylose at A, D-fructose at C and spot the mixture solution at B (Fig. 6.6). Allow the spots to dry. Attach the filter paper to the holder of lid and insert it into the chromatography jar such that the lower end of the paper dips into the solvent while the spots remain above the level of the solvent. The paper should not bend and remain vertically inserted in the jar. Leave the jar undisturbed and allow the solvent to rise till it reaches near the upper end of the paper. Remove the paper from the jar and mark the solvent front immediately with the help of a pencil. Spray the visualizing agent and dry the chromatogram on hot air (oven/ blower/ heat gun) until coloured zones of the spotted sugars become visible. Identify the colored zones and mark their centres with a pencil. Calculate the R_f values and compare the values of individual sugars with those in the mixture to identify them (Fig. 6.7).

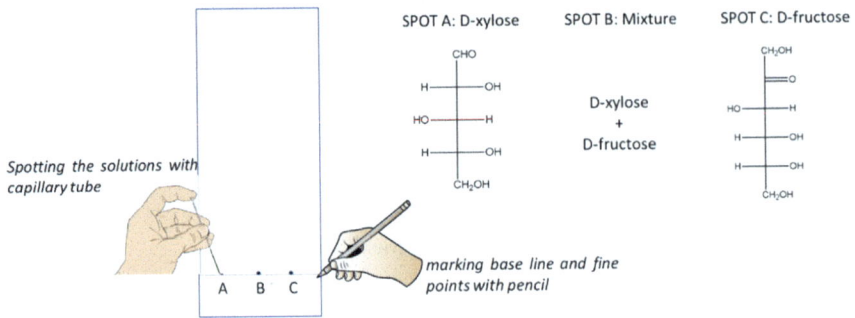

Fig. 6.6 Preparing an ascending paper chromatogram

6.5.2 Separation of Mixture of Amino Acids by Radial Paper Chromatography

In radial or circular paper chromatography the principle for achieving separation of a mixture remains the same as for ascending paper chromatography i.e., partitioning of the solute between stationary and mobile phases. However, the difference is in the direction of flow of the solvent (developing agent). In radial chromatography, the mixture to be separated is spotted at the centre of a horizontally placed paper chromatogram and the solvent flows from the centre towards the periphery thereby bringing separation of the mixture in the form of concentric rings of individual components. Here, the mixture of amino acids will be separated depending upon their individual affinities towards the stationary or mobile phase; which in turn depend upon the structure of the amino acid and the pH of the developing agent. Let us see how.

Amino acids are amphoteric species that contain both acidic (-COOH) and basic (-NH_2) functional groups apart from neutral and hydrophobic functionalities which give different structural features to different amino acids. These structural features in-turn effect the solubility of amino acids in the developing agent. Figure 6.8 illustrates how amino acids exist as positively charged species at low hydrogen ion concentration and negatively charged species at high value of pH, thereby indicating that the pH of the developing agent/mobile phase will largely influence the rate of migration of amino acids giving them unique values of R_f.

Fig. 6.7 Developed ascending paper chromatogram

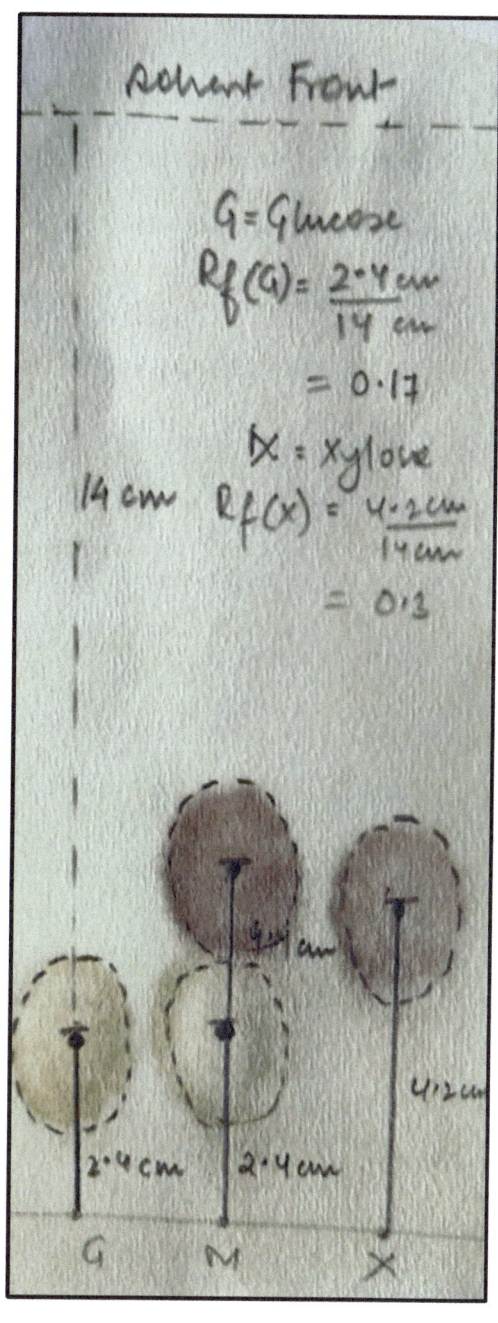

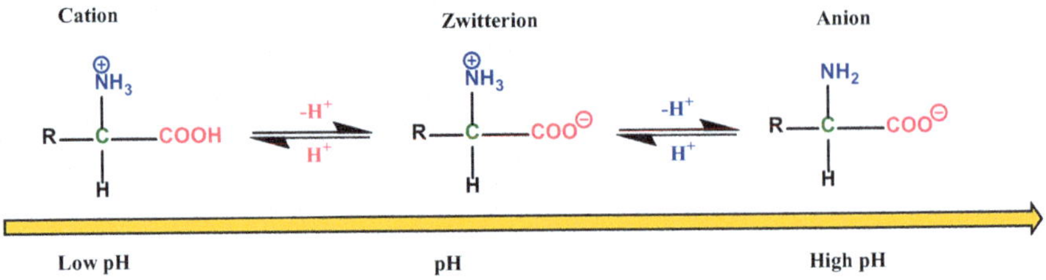

Fig. 6.8 Acid-base behavior of amino acids

6.5.2.1 Chemicals and Apparatus/Equipment Required

Chemicals	Apparatus/Equipment
Solutions of amino acid mixture: glycine and valine (Any other combination of amino acids may be used viz a viz, lysine, phenylalanaine, tyrosine etc.) Developing agent: Butanol, acetic acid and water (48:10:18) Visualizing agent: Ninhydrin solution (prepared by dissolving approximately 1 g of ninhydrin crystals in 100 mL of acetone)	Whatman filter paper no. 1, spotting capillary tubes, petri dishes, cotton wick

6.5.2.2 Experimental Procedure

For the separation of amino acids by radial/horizontal paper chromatography, take requisite amount of solvent (mixture of n-butanol, acetic acid and water abbreviated as BAW) in a petri-dish and cover it with a circular lid so as to allow the chamber to get saturated with the solvent (similar to what was done in the case of ascending paper chromatography). Thereafter, take a circular Whatmann filter paper and mark a point at the centre of this paper with the help of a pencil. Using a capillary, place one drop of the mixture of amino acids at the centre that has been marked in the previous step and carefully dry this. Once the drying of the spot is accomplished, then in a similar manner place another drop at the same marked centre and dry it again. In the next step, prick the centre with the help of an all pin and insert a thin cotton wick (purpose is to facilitate the solvent to travel readily). Place the filter paper with wick on the petri dish so that the cotton wick dips into the solvent. Cover this chromatographic chamber with the lid and leave it undisturbed for a while until the solvent has travelled a certain distance (radially 1 cm before the periphery of the filter paper). Subsequently, take out the filter paper, remove the wick and dry it well. Mark the solvent front with the help of pencil. Using an atomizer, spray ninhydrin

6.5 Experimental Details

reagent onto the filter paper so that the different amino acid components present in the mixture can form complexes with this reagent and the spots can be visualized readily. Dry the chromatogram using either a heat gun or by keeping it in an oven until you are able to see two separate rings of amino acids. Now, mark the rings with the help of a pencil, calculate the R_f values (Fig. 6.9). Developed chromatogram is depicted in Fig. 6.10. The reaction of amino acids with ninhydrin has been depicted in Fig. 6.11.

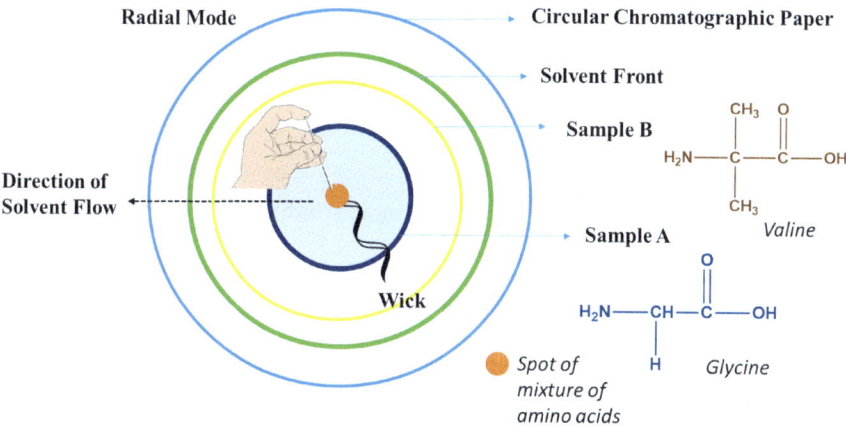

Fig. 6.9 Radial chromatography

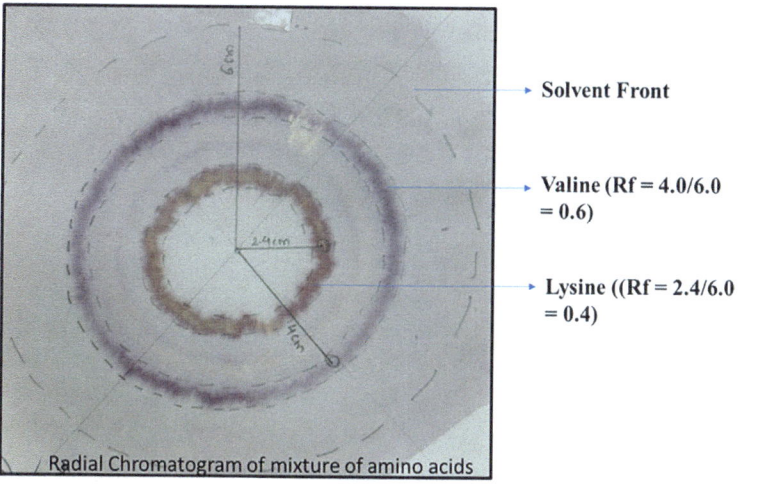

Fig. 6.10 Developed Radial Chromatogram

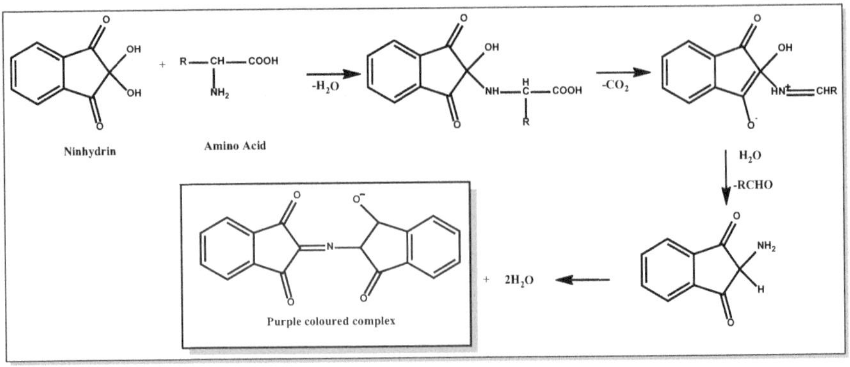

Fig. 6.11 Reaction of amino acids with ninhydrin reagent

6.5.2.3 Tips and Tricks for Radial Chromatography

1. The Whatmann filter paper needs to be carefully held from both the sides.
2. While carrying out the chromatographic separation, the solvent must not travel out of the petri dish boundary.
3. It must be ensured that the petri dish is left undisturbed while the solvent travels or else an efficient separation of the components cannot be established.
4. Always wear latex gloves while working with the ninhydrin reagent spray since this material forms colored compounds with all amino acids, including those in your skin.

6.5.3 Separation of a Mixture of o- and p-Nitrophenol or o- and p-Aminophenol by Thin Layer Chromatography (TLC)

Phenols have a high susceptibility to undergo electrophilic substitution reactions owing to their rich electron density. Nitration provides a good example of one such electrophilic substitution reaction that leads to the generation of a complex mixture containing both *o*- and *p*-nitrophenols (as hydroxyl group is *o*, *p*-directing, hence no meta isomer is formed). Also, on being subjected to reduction, a mixture of *o*- and *p*-amino phenols are formed. In order to separate the mixtures containing *o*- and *p* substituted compounds, it becomes imperative to employ the chromatographic technique.

TLC facilitates the separation of the two components of a mixture with the help of a stationary phase consisting of silica or alumina coated on glass or plastic strip which has been elaborated in detail in the introductory section of this unit. We have learnt that the presence of hydroxyl groups on the surface of either silica or alumina renders it polar. Therefore, when a polar compound is applied to the TLC plate, it undergoes hydrogen bonding interaction with the polar silica/alumina hydroxyl

6.5 Experimental Details

groups. Consequently, more polar a functionality in an analyte is, stronger will be the interaction with the stationary phase. In case of mixture of o and p nitrophenols, the interaction of the silica/alumina will be greater with the p-substituted compound which is more polar. The reason behind its greater polarity as compared to the o-nitrophenol can also be ascribed to the fact that unlike the o-nitrophenol, it does not get involved in intramolecular hydrogen bonding. This implies that the p-nitrophenol will be more strongly bound to the stationary phase and will travel less fast as compared to the o-nitrophenol. The same implication stands true for the o- and p-amino phenols.

6.5.3.1 Chemicals and Apparatus/Equipment Required

Chemicals	Apparatus/Equipment
Solutions of o-nitrophenol and p-nitro phenol or Solutions of o-aminophenol and p-aminophenol Mixture containing both the o and p components Acetone or methanol (could be taken as the solvent for dissolution of these compounds) Developing agent: Ethylacetate, petroleum ether	TLC plate, spotting capillary tubes, chromatography jar with lid and holder UV Chamber

6.5.3.2 Experimental Procedure

Dissolve the given compounds in an appropriate solvent (usually 0.1 g of solute is taken in approximately 5 mL of solvent such as acetone or methanol). Prepare the developing TLC chamber by taking a beaker and filling it with a desired solvent 95:5 ethyl acetate & hexane, till 1 or 2 cm from bottom of the beaker (the mark line should not be dipped into the solution) and cover it with the help of a watch glass to make the container saturated with solvent vapours. Procure a TLC sheet and cut it into a desired size (a rectangular plate of 2 cm × 4 cm) using a pair of scissors to obtain a TLC plate (used for the purpose of chromatographic separation and analysis). Place it on a clean and dry surface keeping the silica coated white part of the plate towards your side. With the help of a pencil, draw a fine horizontal line about 1 cm away from the lower edge of the paper across its width. Mark three points on this base line and label them A, B and C. After this, similar procedure as described in case of separation of sugars may be followed. With the help of different spotting capillaries for each solution, apply the solution of o-nitrophenol at A, p-nitrophenol at C and spot the mixture solution at B (Fig. 6.12). Allow the spots to dry. Thereafter, dip this plate into the TLC chamber and cover it with either a lid or watch glass. Allow the solvent to rise without any disturbance till it reaches 1 cm below the upper end of the TLC plate. Remove the plate from the chamber and immediately mark the solvent front with the help of a pencil, allow it to dry and analyze the distinctive spots under

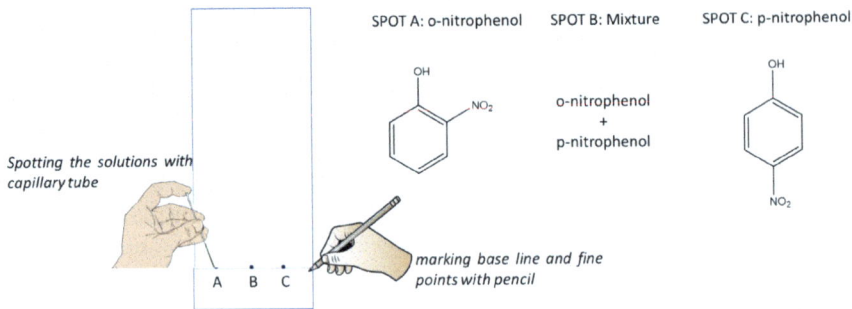

Fig. 6.12 Preparing a TLC plate

Fig. 6.13 Developed TLC plate

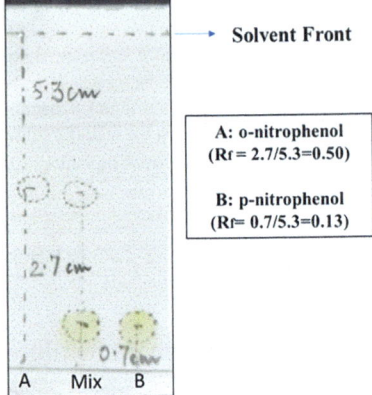

UV chamber. Finally, calculate the R_f value (Fig. 6.13). Similar procedure can be adopted for separation of *o* and *p*-amino phenols.

6.5.3.3 Tips and Tricks for Thin Layer Chromatography

(i) *Remember to fill solvent in the TLC chamber till the level below the pencil mark on the TLC plate.*
(ii) *Use a diluted solution while spotting because if the sample loading exceeds the loading capacity of the TLC plate, then it may appear as a spot with trail or rocket shape spot.*
(iii) *One must allow the solvent to rise without causing any disturbance to the TLC chamber or else it may disturb the flow of solvent front and consequently cause spot with trail.*
(iv) *If a spot does not move during the chromatographic separation process, then the polarity of the solvent may be altered by changing the composition of the solvent.*

6.6 Post-lab Questions

1. In a mixture containing two different sugars (D-Xylose and D-Mannose), which would travel faster and why?

   ....
   ....
   ....
   ....
   ....

2. On what factors does R_f value of a solute depend?

   ....
   ....
   ....
   ....
   ....

3. What is the purpose of using ninhydrin in chromatography of amino acids?

   ....
   ....
   ....
   ....
   ....

4. Which amino acids would have a greater R_f value and why in the following mixtures:
   (i) Aspartic acid and alanine
   (ii) Glutamic acid and glutamine

5. What are the advantages and limitations of paper chromatography?

   ....
   ....
   ....
   ....
   ....

6. What is the prime objective of adding aniline hydrogen phthalate while carrying out the chromatographic analysis of mixture of sugars?

   ....
   ....
   ....
   ....
   ....

7. Amongst o-amino phenol and p-nitrophenol, which would have a greater R_f value and why?

   ....
   ....
   ....
   ....
   ....

8. Why are the spots of certain compounds visible only after irradiation under UV light?

   ....
   ....
   ....
   ....
   ....

9. How would you select a desired solvent for the separation of compounds via TLC?

   ....
   ....
   ....
   ....
   ....

10. What could be the probably reasons of not obtaining a fine spot on the TLC plate during analysis?

    ....
    ....
    ....
    ....
    ....

6.7 Summary Sheet

Separation of sugars using ascending paper chromatography

Mixture of sugars	Distance travelled by component 1	Distance travelled by component 2	Distance travelled by solvent front	R_f value (Exp)

Separation of amino acids using radial paper chromatography

Mixture of amino acids	Distance travelled by component 1	Distance travelled by component 2	Distance travelled by solvent front	R_f value (Exp)

Separation of mixture of o and p-substituted phenols using TLC

Mixture of phenols	Distance travelled by component 1	Distance travelled by component 2	Distance travelled by solvent front	R_f value (Exp)

Results and Discussion

6.7 Summary Sheet

Notes

Chapter 7
Detection of Extra Elements

7.1 Objectives

To identify the extra elements (nitrogen, sulphur and halogens) present in organic compounds.

7.2 About the Experiment

The unit intends to build the foundation behind the need and significance of extra element detection. Through the various tests highlighted comprehensively, you will be able to understand the chemistry behind identification and detection of elements other than carbon and hydrogen that are covalently bonded in organic compounds.

7.3 Introduction

The element carbon which forms the backbone of every organic compound has been known since antiquity to exhibit a unique property known as "catenation" as a result of which it forms covalent bonds with more carbon atoms to form chains or rings. However, organic compounds may contain other elements covalently bonded to carbon. The most commonly encountered elements in organic compounds, other than carbon include halogens (chlorine, bromine, iodine), nitrogen and sulphur and are often termed as 'extra elements'. Apparently, these elements along with carbon bring out scaffold manipulation of heterocyclic motifs that play a crucial role in medicinal chemistry. Nitrogen, appreciably also forms an integral part of naturally occuring proteins, vitamins and hormones that are essential for sustaining life on earth. Primarily, all the organic compounds containing nitrogen are considered as the derivatives of ammonia wherein hydrocarbon radicals substitute one of the hydrogen atoms. Similarly, sulphur is also not only present in a myriad of functionally important organic compounds, but also invariably forms the key component of hormones, enzymes and co-enzymes. Important life-saving antibiotics such as pencillin also contain organosulphur compounds. Interestingly, the compound that imparts a typical flavor to the crushed garlic, known as allicin also comprises of sulphur. Likewise, halogen based organic compounds are widely utilized in industries. In fact, 85% of the pharmaceutical agents rely on halogens such as chlorine, bromine and fluorine to be active. Like herbicides and insecticides, organohalogen compounds are also vital for crop productivity.

The identity of an organic compound is established by carrying out a systematic study of its physical, chemical and spectral properties. Accurate detection of extra elements becomes a critical step in qualitative analysis of organic compound as determination of functional groups present will largely depend upon the extra elements identified. Developed by French chemist, J. L. Lassaigne, the popularly known sodium fusion test or Lassaigne's test is considered to be the most dependable method for detection of extra elements. The test involves fusion of the organic compound being analysed with sodium metal to prepare Lassaigne's extract-a sodium extract. The fundamental reason behind the need for the preparation of sodium extract is that organic compounds contain the extra elements bound covalently to them which need to be converted into their ionic forms for their ready detection. After fusion with sodium, the ionic compounds as obtained can be extracted in aqueous solution and examined by simple chemical tests. In the experimental section, the preparation of Lassaigne's extract is elucidated in detail.

Before we move to the experimental section, let us understand the chemistry behind the sodium fusion test. On reaction with sodium, the extra elements nitrogen, sulphur and halogens present in the organic compounds get converted into corresponding sodium salts i.e. sodium cyanide, sodium sulphide and sodium halides respectively. Sometimes, N and S together react with sodium to form sodium sulphocyanide.

Chemical Reactions involved:

$$Na + C + N \longrightarrow NaCN$$
$$2Na + S \longrightarrow Na_2S$$
$$Na + X \longrightarrow NaX$$
$$Na + C + N + S \longrightarrow NaSCN$$

Once these fused ionic compounds are formed, the qualitative tests of the individual elements can be carried out.

7.3.1 Nitrogen as an Extra Element

Nitrogen is a ubiquitous element that is absolutely essential for driving growth and reproduction in both plants as well as animals. Discovered and isolated long back by a Scottish physician Daniel Rutherford, nitrogen has now become a vital component of thousands of organic compounds (Fig. 7.1).

Many industrially significant organic compounds are used as propellants, explosives and synthetic fertilizers. The significance of this element can be judged from the fact that it has been addressed as life's blueprint for all cells as it is found in all the amino acids that make up proteins and nucleic acids (DNA and RNA), which comprise the hereditary material. The qualitative investigation of presence or absence of nitrogen is of utmost importance as it acts as a diagnostic tool for the characterization of nitrogen based compounds.

Three tests which can be commonly adopted for detecting nitrogen include ferrous sulphate test, benzidine-copper sulphate test and test with *p*-nitrobenzaldehyde (PNB) and *o*-dinitrobenzene (ODNB). All these tests have been covered in detail in Sect. 7.5.3.1.

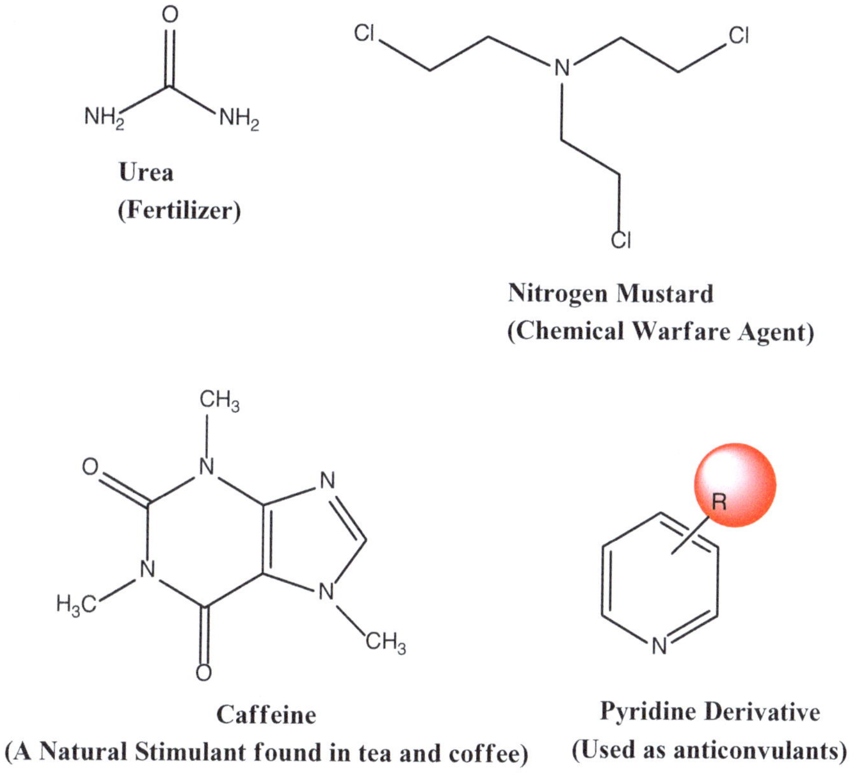

Fig. 7.1 Examples of nitrogen containing organic compounds with their uses

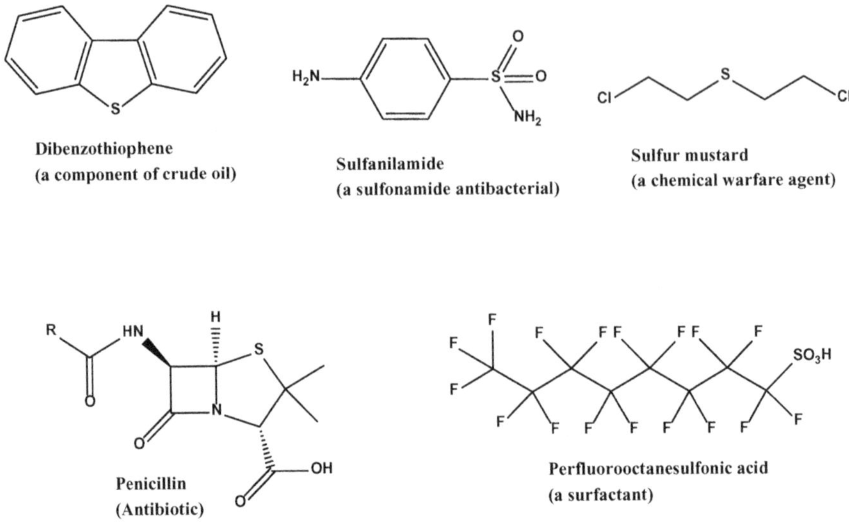

Fig. 7.2 Examples of Sulphur containing organic compounds with their uses

7.3.2 Sulphur as an Extra Element

Sulphur is one of those elements which has been considered absolutely essential for life as nature abounds with organosulphur compounds. As already discussed briefly in the introduction section, we know the significance of organosulphur compounds in the field of drug chemistry, especially the role played by sulfur-containing antibiotics in saving our lives. Apart from supporting our lives, organo sulphur compounds are widely found in fossil fuels such as natural gas and petroleum; the removal of which is of utmost importance for the oil refineries and thus their detection is highly crucial prior to their removal. Structures of some of the sulphur containing organic compounds along with their uses have been depicted in Fig. 7.2.

Since sulphur belongs to the chalcogen group, it possesses properties similar to oxygen, selenium and tellurium and the organo-sulphurs show resemblance to the organo-compounds of these elements. Although there are several tests reported so far for the detection of sulphur, yet this unit will focus on the most frequently employed tests: lead acetate test and sodium nitroprusside test (Sect. 7.5.3.2).

7.3.3 Halogens as Extra Element

Halogen based organic compounds also acquire a special importance in the realm of medicinal chemistry as well as some of the other branches of chemistry (Fig. 7.3). A large number of organo-halogen compounds (especially, organochlorine compounds) are being used as solvents ($CHCl_3$, CH_2Cl_2), pesticides (DDT), flame retardants

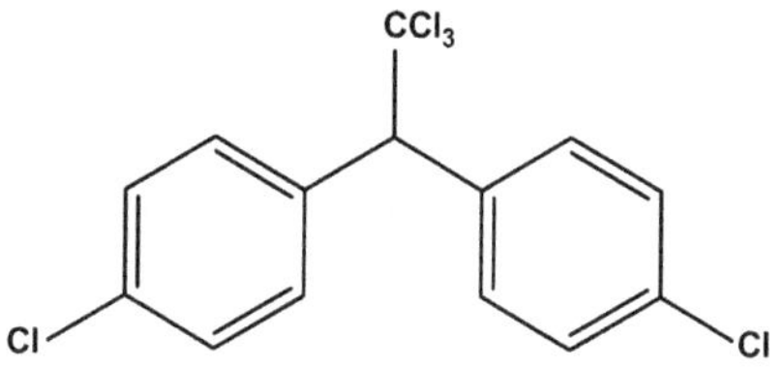

7.3 Introduction

Fig. 7.3 Organo-halogen compounds and their uses

- DDT (An organochlorine pesticide)
- Isoflurane (A general anesthetic) 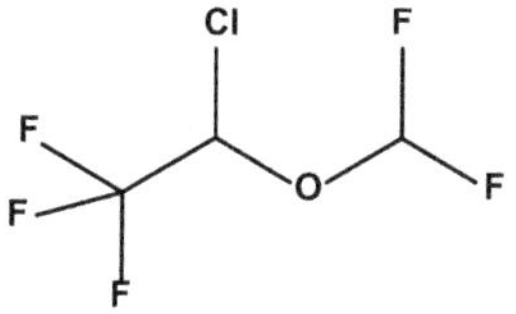
- Tetrabromobisphenol A (TBBPA) (A fire retardant) 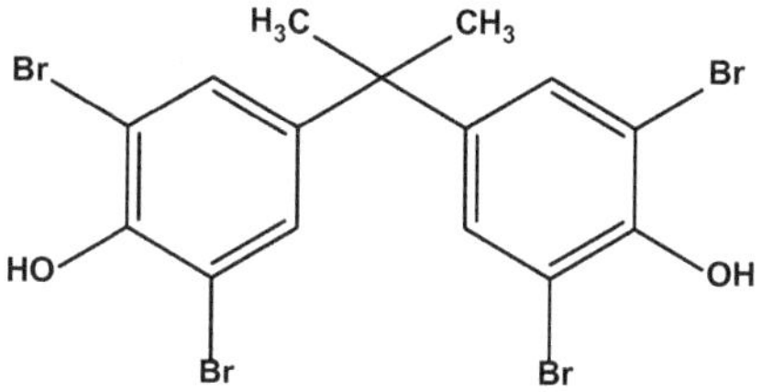
- Ethylene diromide (Antiknock additive in fuels) 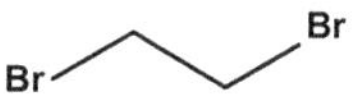

(CF_3Br) and intermediates in the synthesis of dyes (Monastral Fast Green G), drugs (isolflurane) and synthetic polymers like $(C_2F_4)_n$. Interestingly, over two thousand organo-halogen compounds have been identified as naturally occurring materials that are produced by various plants, fungi, bacteria and marine organisms.

The test for halogens is performed after ascertaining the presence or absence of nitrogen and sulphur. Generally, silver nitrate test along with a confirmatory layer test is employed for detection of halogens as extra element. Beilstein test is another method of detection. These tests have been covered in Sect. 7.5.3.4.

Did you know?

Green procedure for detection of extra elements

Lassaigne's method is the conventional method of detection. It has drawbacks like the use of metallic sodium for fusion with organic compounds. Given its reactive nature, handling and disposal of sodium in an undergraduate laboratory could be a cause of concern. In order to overcome the drawbacks of the traditional method of analysis, Middleton pioneered a greener methodology for detection of extra elements – "the sodium carbonate-zinc method" which is also popularly known as the Middleton's method. This is a safer method and possesses a number of advantages such as:- (i) Does not involve the use of highly reactive sodium metal, (ii) Formation of thiocyanate does not occur in case of presence of both N and S since the test for N is not influenced by the presence of S, (iii) Can be employed when Lassaigne's tests do not give satisfactory results especially in case of diazonium salts, azo compounds etc.

7.4 Pre-lab Questions

1. What are extra elements? Explain the significance of the word "extra" in this and why do we need to detect the extra elements accurately in qualitative analysis?

2. What is the purpose of preparing Lassaigne's extract? Give the chemical equations involved. What are the indications of an incomplete fusion process?

3. Give the chemical equations involved for the conversion of N, S and halogens to their ionic forms on fusion with sodium.

4. Why sodium metal is taken to prepare Lassaigne's extract?

7.5 Experimental Section

....
....
....
....
....
....

5. Answer the following:

    (i) Which compound would not give a positive Lassaigne's test for nitrogen?
    - (a) Urea
    - (b) Aniline
    - (c) Isoquinoline
    - (d) Benzoic acid

        Ans:

(b) Which compound does not give a positive result in the Lassaigne's test for nitrogen?

   - (a) Benzamide
   - (b) Urea
   - (c) Aniline
   - (d) Glycerine

6. Mark true or false
   - (a) Hydrazine gives a positive indication for nitrogen in Lassaigne's test.
   - (b) Lassaigne's test is not shown by diazonium salts.

7.5 Experimental Section

7.5.1 Chemicals and Apparatus/Equipment Required

Chemicals	Apparatus/ Equipment
Sodium metal, sodium carbonate, zinc, solid ferrous sulphate, dil. Sulphuric acid, acetic acid, benzidine solution (1% in 50% acetic acid), copper sulphate solution, 0.1 M solution of p-nitrobenzaldehyde, or-dinitrobenzaldehyde in methyl cellosol (2-methoxyethanol), dil NaOH solution, lead acetate solution, sodium nitroprusside, ferric chloride solution, dil. Hydrochloric acid, dil. Nitric acid, silver nitrate solution, conc. Nitric acid, carbon tetrachloride, chloroform	Ignition tubes, china dish, filter paper, boiling tube, test tubes, test tube holder, funnel

7.5.2 Preparation of Lassaigne's Extract

Cut a small piece of sodium metal (approx. 50–100 mg) and dry it by pressing between the folds of the filter paper. Take this freshly cut dried sodium in a clean, dry ignition tube. Heat the tube slowly until the sodium metal forms a shining silvery globule. Remove the ignition tube from the flame and add a small quantity of the organic compound (20–25 mg or a drop in case of liquid) and subject this to heating. It should be noted that the tube is firstly heated on a low flame and subsequently on a stronger flame till it becomes red hot. Keep it in this red-hot condition for a minute or two to ensure that the compound has completely reacted and then plunge this into distilled water (approx. 10 mL) contained in a china dish. Repeat the entire process for one or two more ignition tubes and plunge them again into the same china dish. In case the ignition tube does not break completely, crush it with the help of a glass rod and let the fused material pass into the water. Thereafter, boil the contents for 5 min to reduce the volume to approximately half the initial volume and filter into a boiling tube through a glass funnel using gravity filtration. This filtrate/extract is called as the Lassaigne's extract (Fig. 7.4) which is used for testing the extra elements.

Precautions:

- *Do not touch the sodium with your fingers, handle it with forceps.*
- *There might be slight explosion with compounds such as nitroalkanes, azides, dizonium salts, chloroform and carbon tetrachloride. Therefore, safety goggles must be worn in order to avoid accidents.*
- *Unreacted sodium should be carefully decomposed by reacting with a small amount of ethyl alcohol.*

Did you know?
Why sodium metal is kept under kerosene?

Sodium metal is kept in kerosene to prevent it from reacting with oxygen and moisture in the air. This reaction can be violent and cause a fire. When exposed to air, sodium reacts with oxygen and moisture to form sodium hydroxide. This reaction is highly exothermic, meaning it releases a lot of heat.

7.5.3 Detection of Extra Elements Using Lassaigne's Extract

7.5.3.1 Test for Nitrogen

Ferrous Sulphate Test: Take 1 mL of Lassaigne's extract in a test tube and add freshly prepared ferrous sulphate solution (1–2 mL) or a pinch of $FeSO_4$ (s). This results in the formation of a dirty green precipitate of ferrous hydroxide. Boil the mixture gently for a few seconds and acidify it with dil. sulphuric acid (1–2 mL). Note the

7.5 Experimental Section

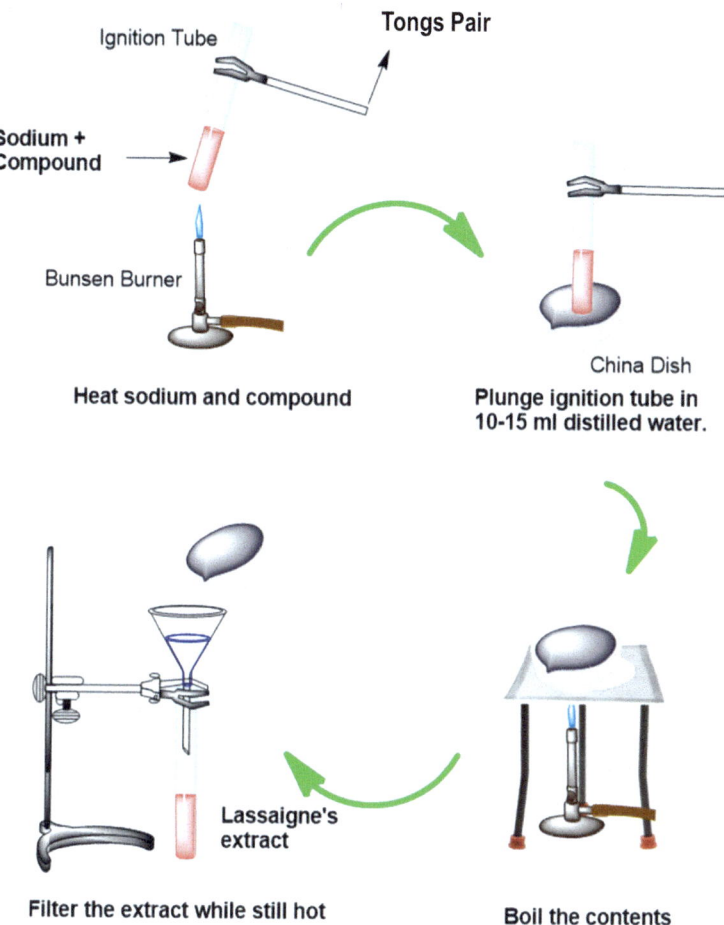

Fig. 7.4 Preparation of Lassaigne's extract

changes. Appearance of Prussian blue colouration confirms the presence of nitrogen as an extra element.

Chemical Reactions involved:

$$6\ NaCN + FeSO_4 \longrightarrow Na_4[Fe(CN)_6] + Na_2SO_4$$
Sodium Ferrocyanide

$$3Na_4[Fe(CN)_6] + 2Fe_2(SO_4)_3 \longrightarrow Fe_4[Fe(CN)_6]_3 + 6\ Na_2SO_4$$
Ferri-Ferrocyanide
Prussian Blue

Benzidine Copper Sulphate Test: To about 1 mL of Lassaigne's extract, add 3–4 drops of acetic acid to acidify it. Thereafter, add freshly prepared benzidine solution (2–3 drops, 1% in 50% acetic acid) and shake this mixture. Afterwards, add copper sulphate solution (1–2 drops, 1%) and observe the changes. Formation of a blue precipitate or a blue colouration indicates the presence of nitrogen.

Chemical Reactions involved:

$$CuSO_4 + 2NaCN \longrightarrow Na_2SO_4 + Cu(CN)_2$$

$$2Cu(CN)_2 + 2\,H_2N\text{-}C_6H_4\text{-}C_6H_4\text{-}NH_2 \longrightarrow \text{Copper Cyanobenzidine Complex (Blue)}$$

Test with PNB and ODNB: To 1 ml of Lassaigne's extract, add few drops of dil NaOH in order to make it alkaline and then add *p*-nitrobenzaldehyde (PNB) solution (4–5 drops) followed by *o*-dinitrobenzaldehyde (ODNB) solution (4–5 drops). The appearance of a purple colouration indicates the presence of nitrogen.

Chemical Reactions involved:

[A] Purple

7.5 Experimental Section

7.5.3.2 Test for Sulphur

Lead Acetate Test: Take 1 mL of the Lassaigne's extract and acidify it with dil acetic acid (2 mL approx.) To this, add 4–5 drops of lead acetate solution. In case, sulphur is present, a black precipitate is obtained which is due to the formation of lead sulphide (PbS).

Chemical Reaction involved:

$$Na_2S + (CH_3COO)_2Pb \longrightarrow PbS\downarrow + 2CH_3COONa$$
$$\text{Lead Sulphide}$$
$$\text{(Black ppt)}$$

Sodium nitroprusside test: To 1 mL of Lassaigne's extract, add 1–2 drops of freshly prepared sodium nitroprusside solution. The appearance of a deep reddish violet colouration indicates the presence of sulphur.

Chemical Reaction involved:

$$Na_2S + Na_2[Fe(CN)_5NO] \longrightarrow Na_4[Fe(CN)_5NOS]$$
$$\text{Sodium nitroprusside} \qquad \text{Sodium sulphonitroprusside}$$
$$\text{(Reddish Violet)}$$

7.5.3.3 Tests for Nitrogen and Sulphur (When Present Together)

It may so happen that both nitrogen and sulphur are present as extra elements in an organic compound. In such a case, their presence may be detected through the following tests.

$FeCl_3$ Test: To 1 mL of Lassaigne's extract, add 5–6 drops of dilute HCl followed by 2–3 drops of ferric chloride solution. Appearance of blood red colouration indicates the presence of sulphur and nitrogen present together in the compound.

Chemical Reaction involved:

$$3NaSCN + FeCl_3 \longrightarrow Fe(SCN)_3 + 3NaCl$$
$$\text{Ferric thiocyanate}$$
$$\text{(Blood Red)}$$

Cobalt Nitrate Test: Take 1 mL of Lassaigne's extract and add 1 mL of dilute HCl in order to acidify it. Then, add 1 mL of alcohol (ethanol) followed by 4–5 drops of cobalt nitrate solution. A blue colouration indicates the presence of both nitrogen as well as sulphur.

Chemical Reactions involved:

$$NaSCN + HCl \longrightarrow HSCN + NaCl$$
$$4\ HSCN + Co(NO_3)_2 \longrightarrow H_2Co(SCN)_4 + 2\ HNO_3$$
$$\text{Cobaltithiocyanic acid (blue)}$$

7.5.3.4 Test for Halogens

Case (i) If nitrogen is not present

Silver Nitrate Test: Add 2–3 drops of dilute nitric acid to about 1 mL of Lassaigne's extract. Add 0.5–1 mL of silver nitrate solution to this. Appearance of a white precipitate soluble in NH_4OH indicates presence of chloride (Scheme 3). A pale yellow precipitate sparingly soluble in NH_4OH or soluble in excess of NH_4OH indicates the presence of bromide while a yellow ppt insoluble in this ammonia solution indicates the presence of iodide (Fig. 7.5).

Case (ii) If nitrogen or/and sulphur is present

In case, an organic compound consists of nitrogen and sulphur, the sodium cyanide or sodium sulphide formed during fusion may interfere with the analysis of halogens by forming a white precipitate of silver cyanide or black precipitate of silver sulphide. Therefore, before proceeding further, the extract should be boiled with conc HNO_3 in order to decompose the sodium cyanide and or sodium sulphide and expel them in the form of HCN and H_2S gases. Perform the test as follows:

Boil 1 mL of lassaigne's extract with 2–3 mL of conc. HNO_3 for about 5–7 min followed by addition of 1 mL of $AgNO_3$ solution. Observe the changes as in silver nitrate test.

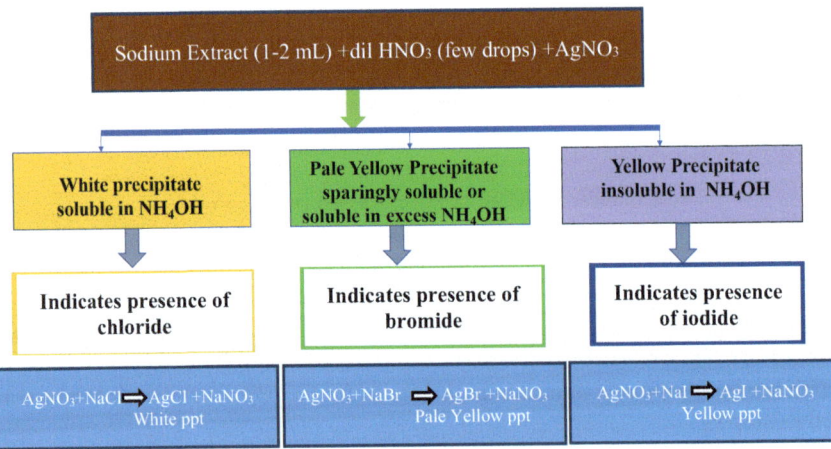

Fig. 7.5 $AgNO_3$ test for halides

7.5 Experimental Section

Chemical Reactions involved:

Interference by silver cyanide and silver sulphide

$$NaCN + AgNO_3 \longrightarrow AgCN\downarrow + NaNO_3$$
$$\text{White ppt}$$
$$Na_2S + 2AgNO_3 \longrightarrow Ag_2S\downarrow + 2NaNO_3$$

Removal of cyanide and sulphide as gases

$$NaCN + \text{conc. } HNO_3 \longrightarrow HCN(g)\uparrow + NaNO_3$$
$$Na_2S + \text{conc. } HNO_3 \longrightarrow H_2S(g)\uparrow + 2NaNO_3$$

Layer Test: Perform this test only when the silver nitrate test is positive. To 2 mL of Lassaigne's extract, add conc HNO_3 (1 mL) and warm this mixture for few seconds. Then add, CCl_4 or $CHCl_3$ (1 mL) and shake well. If the organic layer appears violet then presence of iodide is indicated or else if the layer retains an orange-brown colour, then the presence of bromide is confirmed. A colourless organic layer indicated the presence of chlorine. The oxidant oxidizes sodium halide (chloride, bromide or iodide) to free halogens (chlorine, bromine or iodine) which gives characteristic colour in organic layer.

Chemical Reactions involved:

$$2NaCl + HNO_3 \xrightarrow{[O]} 2NaNO_3 + Cl_2\downarrow + H_2O$$
$$\text{Colourless organic layer}$$

$$2NaBr + HNO_3 \xrightarrow{[O]} 2NaNO_3 + Br_2\downarrow + H_2O$$
$$\text{Orange or brown organic layer}$$

$$2NaI + 2HNO_3 \xrightarrow{[O]} 2NaNO_3 + I_2\downarrow + H_2O$$
$$\text{Violet organic layer}$$

Beilstein Test: Take a copper wire and heat a small loop of this wire in a flame until the flame is no longer coloured. Then cool this wire, dip it in the organic compound and heat it again in the edge of a clear flame. A green colouration indicates the presence of halogens.

Note: -Fluorine is not responsive in this test. It is not very adsivable to follow the Beilstein's test as it may fail for volatile liquids, as they evaporate even before the copper wire is heated sufficiently in order to cause the decomposition of the given compound. Also, certain compounds not containing halogen may also impart green colouration. Examples of such compounds include pyridine, organic acids, urea and quinoline derivatives.

-Ensure that the extract with nitric acid is not heated for a longer time. In case of longer heat subjection, the halogen formed due to the oxidation of sodium halide may escape.

7.6 Green Method for Detection of Extra Elements (Middleton's Method)

The present practice of using metallic sodium for fusion with organic compound is a cause of serious concern because of the hazards associated with using sodium. This method involves heating up of an organic compound with a mixture of sodium carbonate and zinc dust wherein the nitrogen and halogens get converted into sodium cyanide and sodium halide respectively, while the sulphur is precipitated in the form of zinc sulphide. Thereafter, the water soluble cyanides and halides can be detected conventionally as done in the Lassaigne's method, while on the other hand the insoluble sulphide is subjected to an acid treatment which results in the evolution of hydrogen sulphide which can be detected by reacting with lead acetate.

7.6.1 Preparation of Fusion Extract for Middleton's Method

To prepare the zinc-alkali fusion extract by green method, take zinc dust (0.2 g) and sodium carbonate (0.3 g) and mix thoroughly with the organic compound (approx. 0.05 g) in an ignition tube and heat gently followed by strong heating until the tube becomes red hot. Plunge the ignition tube into a china dish containing approx. 10 mL distilled water. The fusions are repeated using at least 2–3 ignition tubes and boil to reduce the volume to half to get a concentrated fusion extract. Filter the extract into a boiling tube using gravity filtration and collect the residue in another china dish to detect the extra elements.

7.6.2 Detection of Extra Elements by Middleton's Method

The extract prepared by Middleton's method contains dissolved cyanides and halides, while sulphur is precipitated in the form of zinc sulphide. Use the filtrate (as described in subsection 7.6.1) to carry out the detection of nitrogen and halogen, the same way as done in Lassaigne's method (section 7.5.2). Take the residue in a china dish and acidify it with 5–6 mL of dilute hydrochloric acid. Cover the china dish immediately with a filter paper moistened with lead acetate solution. If sulphur is present, H_2S evolved will turn the lead acetate paper deep brown or black due to the formation of lead sulphide. The detection tests are summarized in tabular form as given in Table 7.1.

Green Context

- Through the greener alternative procedure suggested, the risk of explosion and fire hazard which are often met while carrying out the same experiments using metallic sodium can be completely eliminated.

7.7 Post-lab Questions

Table 7.1 Green method for detection of extra elements

Sl. no	Experiment	Observation	Inference
1	Take 1 mL of fusion extract in a test tube and add a pinch of $FeSO_4$ followed by subjecting the contents to heating. Cool and add few drops of dil. H_2SO_4	Prussian Blue Coloration	N present
2	Take the residue obtained while preparing fusion extract in a china dish and acidify it with 5–6 mL of dilute hydrochloric acid. Cover the china dish immediately with a filter paper moistened with lead acetate solution.	Black/deep brown ppt on filter paper	S present
3	To 1 mL of fusion extract, add 1–2 mL of $FeCl_3$ solution	Blood Red Colouration	N and S both present
4	Boil 1 mL of fusion extract with few drops of conc HNO_3 boil. Cool and add 1–2 mL of $AgNO_3$ solution. Layer test can also be performed for further confirming the halogens (subsection 7.5.3.4)	Curdy white ppt (Soluble in NH_4OH) Pale yellow ppt (Partly soluble in NH_4OH) Yellow ppt (Insoluble in NH_4OH)	Chlorine present Bromine present Iodine present

Note: This can be conducted as an extension experiment for the undergraduate students so that they acquire skills beyond the curriculum and are motivated to think critically and understand the problems associated with the traditional practices and the need to develop alternative methodologies

7.7 Post-lab Questions

1. Why do we need to dry the sodium metal before subjecting it to fusion?

   ....
   ....
   ....
   ....
   ....
   ....

2. Why should we never touch sodium with our fingers and rather use forceps?

   ....
   ....
   ....
   ....
   ....
   ....

3. How is the sodium metal usually stored?

4. Why distilled water and not tap water is used in the preparation of Lassaigne's extract?

5. Why is the Lassaigne's extract alkaline?

6. Why do you need to use a freshly prepared solution of $FeSO_4$ while testing nitrogen?

7. When nitrogen and/or sulphur are present, why do you need to acidify the extract with conc. nitric acid for testing the presence of halogens?

7.7 Post-lab Questions

8. What leads to the colour in organic layer, while performing the test for halogens?

   ....
   ....
   ....
   ....
   ....
   ....

9. While performing the layer test, why should we not heat the extract with nitric acid for a longer time?

   ....
   ....
   ....
   ....
   ....
   ....

10. Why do you think Beilstein's test alone is not enough to detect halogens?

    ....
    ....
    ....
    ....
    ....
    ....

11. Why does hydrazine hydrate not give a positive test for nitrogen as an extra element? Would addition of sodium carbonate while preparing the fusion extract result in a positive test for nitrogen in this case?

    ....
    ....
    ....
    ....
    ....
    ....

12. Discuss the advantages of green method for detection of extra elements.

    ....
    ....
    ....
    ....
    ....
    ....

13. How would you test sulphur using Middleton's method?

   ....
   ....
   ....
   ....
   ....
   ....

14. Give the answer to the following:
    (a) When sodium nitroprusside is added to Lassaigne's extract, a violet colour is formed. What does this indicate?

   ....
   ....

    (b) In the Lassaigne's test for the detection of nitrogen in an organic compound, the prussian blue colour is due to the formation of which compound?

   ....
   ....

7.8 Summary Sheet

Entry	Experiment	Observation	Inference
1.	Test for Nitrogen		
2.	Test for Sulphur		
3.	Test for Nitrogen and Sulphur (present together)		
4.	Test for Halogens		

Results and Discussion

....
....

Notes

7.8 Summary Sheet

Viva-Voce Questions with Answers

Q.1 What is the purpose of using fluted filter paper in filtration?

Ans. Fluted filter paper plays a significant role in increasing the surface area and henceforth the rate of filtration.

Q.2 Mention the technique that could be utilized for the purification of the following substances

(a) **A mixture of compounds containing camphor and common salt**
(b) **A mixture of o-nitrophenol and p-nitrophenol**
(c) **A liquid containing non-volatile impurities**

Ans. (a) Sublimation, (b) Steam Distillation (o-nitrophenol being volatile distills over along with water while p-nitrophenol being non-volatile remains in the flask) and (c) Simple Distillation (non-volatile impurities will remain as residue in the flask).

Q.3 Which bath liquids can be used instead of concentrated sulphuric acid for determination of melting point?

Ans. Paraffin oil and silicone oil

Q.4 What happens to the boiling point of water on addition of NaCl?

Ans. With the addition of soluble impurities, boiling point increases.

Q.5 Why should a sample be crushed well to form a powder before melting point is determined?

Ans. To increase the surface area that would help in uniform, rapid and more efficient transfer of heat.

Q.6 Can we use calcium and potassium in the preparation of Lassaigne's extract instead of sodium?

Ans. No, potassium cannot be used due to its high reactivity which may cause violent reaction; while calcium has low reactivity as compared to sodium making it unsuitable for converting covalently bonded extra elements to their ionic forms.

Q.7 In the test for detection of nitrogen, can Mohr's salt be used instead of ferrous sulphate?

Ans. Yes, as the basic purpose is to furnish ferrous ions which are present in Mohr's salt also.

Q.8 Why is freshly prepared ferrous sulphate solution recommended in the test for nitrogen?

Ans. This is because, ferrous sulphate undergoes hydrolysis, forming pale yellow coloured Fe^{3+} which interferes with the prussian blue colour obtained in the test.

Q.9 What is the need of boiling an alkaline solution containing ferrous ions during the test for nitrogen?

Ans. Boiling is necessary so as to convert/oxidize ferrous ions into ferric ions which then react with sodium hexacyanoferrate(II) to generate ferri-ferrocyanide, responsible for imparting the prussian blue colouration that confirms the presence of nitrogen.

Q.10 Which of the following nitrogen based compounds do not respond to the Lassaigne's test?

(a) **Hydrazine**
(b) **Urea**
(c) **Diazonium Salt**

Ans. (a) and (c) do not respond to the Lassaigne's test.

Hydrazine does not respond to the Lassaigne's test due to absence of carbon. However, if activate charcoal is added to the compound, it will give the positive test for nitrogen.

Nitrogen is lost from diazonium salts on heating, before it can react with sodium metal, hence they do not show positive LL assaigne's test.

Q.11 Why do we need to perform blank test, while testing for chloride?

Ans. This is because tap water may contain a considerable amount of chlorine which can give a false positive test for chloride even if the compound does not contain chloride.

Q.12 Why is acidification with nitric acid done before addition of silver nitrate in the test for halogens?

Ans. This is done in order to prevent the precipitation of silver hydroxide or oxide as the Lassaigne's extract is alkaline in nature.

Q.13 Why is it necessary to boil the filtrate with conc nitric acid if nitrogen and/or sulphur is/are present before performing the test for halogens?

Ans. To expel cyanide and sulphide as HCN and H_2S, thereby preventing the precipitation of silver cyanide or silver sulphide which interfere with detection of halogens.

Q.14 The Rf values obtained on a paper chromatogram for a mixture containing compounds A and B are 0.45 and 0.6 respectively. Which of the two compounds is more polar?

Ans. Compound A is more polar as it travels a shorter distance and hence is more strongly adsorbed over the chromatogram.

References

Ahluwalia VK, Aggarwal R (2001) Comprehensive practical organic chemistry: preparation and quantitative analysis. Universities Press

Furniss BS (ed) (2011) Vogel's textbook of practical organic chemistry. Pearson Education India

Gattermann L (1909) The Practical Methods of Organic Chemistry

Leonard J, Lygo B, Procter G (2013) Advanced practical organic chemistry. CRC Press

Mann FG, Saunders BC (1975) Practical organic chemistry. Orient Blackswan

Pavia DL, Kriz GS, Lampman GM, Engel RG (2015) A small scale approach to organic laboratory techniques. Nelson Education

Reid RE (1927) Handbook of organic analysis, qualitative and quantitative. Clarke, Hans Thacher

Sharp JT (ed) (2012) Practical organic chemistry: a student handbook of techniques. Springer Science & Business Media

Vogel AI (1956) Practical organic chemistry. Longmans 2:676–681

The manufacturer's authorised representative in the EU is Springer Nature Customer Service Centre GmbH, Europaplatz 3, 69115 Heidelberg, Germany. If you have any concerns regarding our products, please contact ProductSafety@springernature.com

Printed and bound by CPI Group (UK) Ltd, Croydon, CR0 4YY

15/12/2025

02019692-0002